Contesting Water Rights

Mangala Subramaniam

Contesting Water Rights

Local, State, and Global Struggles

Mangala Subramaniam
Department of Sociology
Purdue University
West Lafayette, IN, USA

ISBN 978-3-030-09039-5   ISBN 978-3-319-74627-2 (eBook)
https://doi.org/10.1007/978-3-319-74627-2

© The Editor(s) (if applicable) and The Author(s) 2018
Softcover re-print of the Hardcover 1st edition 2018
This work is subject to copyright. All rights are solely and exclusively licensed by the Publisher, whether the whole or part of the material is concerned, specifically the rights of translation, reprinting, reuse of illustrations, recitation, broadcasting, reproduction on microfilms or in any other physical way, and transmission or information storage and retrieval, electronic adaptation, computer software, or by similar or dissimilar methodology now known or hereafter developed.
The use of general descriptive names, registered names, trademarks, service marks, etc. in this publication does not imply, even in the absence of a specific statement, that such names are exempt from the relevant protective laws and regulations and therefore free for general use.
The publisher, the authors and the editors are safe to assume that the advice and information in this book are believed to be true and accurate at the date of publication. Neither the publisher nor the authors or the editors give a warranty, express or implied, with respect to the material contained herein or for any errors or omissions that may have been made. The publisher remains neutral with regard to jurisdictional claims in published maps and institutional affiliations.

Cover image: ("Water and People," 2017) © S. Brintha Lakshmi

Printed on acid-free paper

This Palgrave Macmillan imprint is published by the registered company Springer International Publishing AG part of Springer Nature
The registered company address is: Gewerbestrasse 11, 6330 Cham, Switzerland

Preface

Water is a basic need for all living beings. Yet access to clean water for all people across the world has been a challenge. Not having sufficient and safe water means being susceptible to diseases. It also leads to loss of time and educational and employment opportunities. Low incomes and limited access to water also mean choosing between paying for water, food, school fees, or medicines. According to the World Health Organization (WHO) and UNICEF, 748 million people around the world lack access to an improved (clean, uncontaminated) drinking water source, while billions more lack drinking water that is really safe.

Water as an issue has been constantly visible in my travels for my field-based research, especially in rural areas, for more than a decade now. I carry bottled water as I always worry about safe water in the regions to which I travel. And then I feel compelled to explain to my interviewees why I carry bottled water, especially because it could be construed as adhering to caste norms. This is ironic, considering that research about control over and privatization of water resources is the focus of this book.

This book is an outcome of research I began way back in 2006, when I became interested in environmental resources and justice issues. Being fortunate to receive a competitive Asian Initiative Research grant from Purdue University, I used my one-semester sabbatical in 2007 to travel through parts of Rajasthan—Udaipur and Alwar districts. Rajasthan, located in the north-western part of India, is relatively dry and infertile. The area includes some of the Thar Desert, also known as the Great

Indian Desert. I had learnt of ongoing development work in Udaipur and with the assistance of Sewa Mandir (a non-governmental organization (NGO)), I had the opportunity to visit interior rural areas and see at close hand the challenges posed by the shortage of water as well as the indigenous water saving techniques used. People had told me about the 'pani sansad' (water parliament) sessions in the Alwar district and I was curious to attend it. Not knowing anyone at Tarun Bharat Sangh (TBS), an NGO that enabled mobilizing residents for addressing the water crisis and protecting the environment, I faced some difficulties in locating the village, but was warmly welcomed at the session. As I observed the vibrant discussions about water-related issues among many of the men who approached the mike placed at the front of the gathering, I noted that women, with their heads covered by their saris, sat in silence. The image starkly brought the gender differences in water-related decision-making and management to the forefront.

Subsequently in 2008, I was contacted by the Program Officer of the Ratan Tata Trust to inform me of a project they were interested in funding in collaboration with the International Water Management Institute (TATA-IWMI). It was about social movements concerning water. The core research group included a representative from TATA-IWMI, the Society for Promoting Participative Ecosystem Management (SOPPECOM), the Madras Institute of Development Studies, India, and myself. The grant was released by the Trust. This was a tremendous opportunity to gain insights into the various small- and large-scale efforts to raise awareness about managing water resources and enable community ownership of such resources.

I continued to maintain an interest in the struggles about water and published two articles about water rights (in 2012 and 2014). My interest in the challenges to the water crisis was revived and sustained through the discussions among a group of faculty members and graduate students focused on social movements (at Purdue). Late in 2013, I was involved in a committee that put together a list of scholars for the international workshop on Environmental Justice and Equitable Access to Natural Resources organized by Purdue's Center for the Environment. A principal concern of the workshop was to bring together scholars, activists, and others engaged in interdisciplinary environmental justice work.

And then in 2016, when the water crisis unfolded in Flint, Michigan, I noticed some similarities between the developed and developing world relating to access to water. I wrote a blog as a member of the editorial board of the journal *Gender & Society*. My concerns were with the ways in which the marginalized—the poor, racial minorities, and women—are rarely recognized and included in discussions about managing environmental resources such as water. That was when I began to seriously consider a book about water—not only focusing on local struggles, but spanning the local and global levels, and how such struggles need to be understood by considering the social, political, and economic context in which they occur.

As Palgrave Macmillan had contacted me a few years earlier about the possibility of pursuing a book through them, I approached them with the idea. Their quick response and interest led to a formal contract for this book.

West Lafayette, IN, USA   Mangala Subramaniam
August 2017

ACKNOWLEDGEMENTS

I acknowledge the countless NGOs, groups, activists, and scholars in India and elsewhere who welcomed me to see their work related to water and other development activities. It has been an incredible way to gain knowledge and insights into concerns about water. I am grateful to Purdue University for the Asian Initiative Research grant and to TATA-IWMI for the opportunities to meet activists and learn of water-related projects across India. I also would like to thank Neelima Khetan, Sanjiv Phansalkar, the employees of Tarun Bharat Sangh, and the many village residents in the Udaipur and Alwar districts. I thank the many activists and NGO representatives who shared information about their projects as part of the TATA-IWMI grant-related research.

I have benefited from the research assistance of students. Jared Wright downloaded many documents pertaining to the global water forums as part of the work for the social movements group. I continued to track the documents and downloaded new ones, which I analyze in the book.

My sincere thanks to Alexis Nelson of Palgrave Macmillan for working with me as I put together a book proposal. Her close review of my proposal, and particularly the book content, led to very useful comments for me to revise and sharpen it. This resulted in relatively good reviews for the book. After her departure, I worked with Kyra Saniewski and then later with Mary Al-Sayed, the new editor. I am very grateful to both Mary and Kyra for being exceptionally patient with me as I completed the manuscript. I delayed the submission of my manuscript for personal reasons and especially as I began a new position as the Butler Chair and

Director of the Butler Center. Thanks are also due to the publisher for making the watercolor illustration work for the cover possible.

I am thankful for the help and support of family and friends outside my immediate academic realm. My late father, P. R. Subramanian, and late mother, Narayani, were very supportive of my research and writing endeavors. My mother was a feminist in her own way. She celebrated my accolades and constantly encouraged me to strive for more. I am indebted to my parents for their tremendous ability to make anything possible for me. I am very thankful for the support of my siblings and other family members, Vasanta, Brintha, Shobha, Ravi, and Yogi, who have all contributed to making this book project possible. Without their help and boundless energy, this book would not have been completed. I also appreciate Brintha's enthusiasm to develop an illustration for the cover of this book.

Contents

1 Introduction: Water Crisis   1

2 Neoliberalism, the Ambivalent State, and Community Struggles   25

3 Contesting Water Rights from "Below"   59

4 Controlling Water Resources from "Above": Global Water Forums   95

5 Conclusion   153

Glossary   169

Index   173

LIST OF TABLES

Table 3.1	Main indicators for Kerala and Rajasthan	63
Table 3.2	Comparison of cases	67
Table 3.3	Comparison: struggles, state, strategies	80
Table 4.1	World Water Forums	102
Table 4.2	Alternative World Water Forums	112

CHAPTER 1

Introduction: Water Crisis

The recent and ongoing water crisis in Flint, Michigan draws attention to an issue that has affected and continues to affect people across the world: access to clean/safe water. This crisis began when the source of water supply to Flint was changed from Lake Huron and the Detroit River to the Flint River to save money. Among those affected are the marginalized and particularly the poor (Subramaniam 2016). The right to clean water is a demand of people not only in Flint, but of people across the world. To make sense of the water crisis, scholars and policy makers have considered the role of the government in promoting privatization directly or indirectly, the economic costs and benefits of promoting efficient systems of managing water use and allocation, and the role of the community that represents race and class differences in negotiating and resisting the privatization of water resources that are largely perceived by the community as common property resources (Olivera 2014; Moyo and Liebenberg 2015; Subramaniam 2014; Mullin 2014; Breakfast et al. 2014; Johnson et al. 2016, among others). Communities struggle as ground water levels deplete and the contamination of water adversely affects the lives of people, particularly, marginalized people.

In this chapter, I provide an overview of the water crisis and also include a discussion of the early approaches to studies about water. I cover three main issues. First, among the environmental specters confronting humanity in the twenty-first century, shortage of water is at the top of the list. Providing details of the estimates of water availability and uses will provide a reader with an overview of the landscape of concerns with water: an overview of the world's water resources. I incorporate details of depletion and management of water resources as reported by various institutions such as the United Nations, the World Resources Institute and the World Bank among others. Second, the water crisis—and particularly access to and control over water resources—spans across the developed and developing world. Concerns about access to clean water and citizens' control over water are unique to a specific locale or region of the world (Loftis 2016; Subramaniam and Zanotti 2015; Akiwumi 2015; Boelens and Zwarteveen 2005; Olivera 2004; Snitow and Kaufman with Fox 2007). Third, those most affected by the shifts in the control over water resources and the privatization of water supply mechanisms are the marginalized and the poor. In addition to a review of past work, I include an overview of the three cases and transnational forums that are examined in the following chapters. This is followed by the sources and types—primary and secondary—of data that I use for the analysis. In the last section of this chapter, I provide an overview of the following chapters in this book.

WATER RESOURCES: AVAILABILITY AND USE OF WATER

Among the environmental specters confronting humanity in the twenty-first century, shortage of water is at the top of the list, particularly in the developing world. According to the World Resources Institute, more than a billion people currently live in water-scarce regions and as many as 3.5 billion could experience water scarcity by 2025. Moreover, if consumption patterns continue at current rates, 2.7 billion people would face severe water shortages by 2025 (Montainge 2002). Increasing pollution degrades freshwater and coastal aquatic ecosystems.[1] And climate change is poised to shift precipitation patterns and speed glacial melt, altering water supplies and intensifying floods and drought. The bleak future is also attributed to the projected growth of the global population, from more than six billion today to an estimated nine billion in 2050. Yet the amount of fresh water on earth is not increasing.

Several countries around the world are facing a severe water crisis. Water-abundant regions have become water-scarce, and water-scarce regions face water famines. It is important to consider both supply and demand factors, combined with the effects of neoliberal policies, in order to understand the water crisis. While knowledge of patterns of water use is scattered, it is estimated that about 4000 cubic kilometers of fresh water are used each year globally. Irrigation for agriculture accounts for about 70% of this usage, industrial use consumes 20%, and domestic use accounts for the remaining 10%. About 20% of this water is drawn from groundwater sources, the use of which increased fivefold in the twentieth century (World Water Assessment Programme 2009).

Groundwater plays a substantial role in water supply, ecosystem functioning and human well-being. Worldwide, 2.5 billion people depend solely on groundwater resources to satisfy their basic daily water needs, and hundreds of millions of farmers rely on groundwater to sustain their livelihoods and to contribute to the food security of so many others (UNESCO 2012). Groundwater reportedly provides drinking water to at least 50% of the global population and accounts for 43% of all water used for irrigation (Groundwater Governance n.d.). Groundwater also sustains the base flows of rivers and important aquatic ecosystems. Uncertainty over the availability of groundwater resources and their replenishment rates pose a serious challenge to their management and, in particular, to their ability to serve as a buffer to offset periods of surface water scarcity (Van der Gun 2012). Overuse of groundwater resources has led to a rapid drawdown of aquifers in some areas, and the sustainability of such usage has become the focus of much international attention (World Water Assessment Programme 2009).

Overpumping of aquifers and falling water tables have affected northern China, India, Algeria, Egypt, Iran, Mexico, and Pakistan, as well as California in the United States, also face major water shortages (World Water Assessment Programme 2009). The overpumping of aquifers may be for agricultural or municipal use. US farmers are withdrawing water from the Ogallala Aquifer, which underlies the Great Plains, at an unsustainable rate, with a third of the Texas portion already significantly depleted. The water table under the North China Plain, which produces about half of China's wheat and corn, is steadily dropping (Montaigne 2002).

Water use patterns vary by region. In Africa, most of Asia, Oceania, Latin America, and the Caribbean, agriculture accounts for the majority of usage, while in Europe and North America, industry and energy-related usage are far more significant. In 2006, 87% of the world's population had access to improved drinking water sources, leaving Sub-Saharan Africa and Oceania as the only regions not on target for the United Nations' Millennium Development Goals drinking water recommendations. However, this amounts to more than 800 million people without access to suitable drinking water and if current trends continue, 2.4 billion people will still be without access to basic sanitation in 2015 (World Water Assessment Programme 2009).

As the population continues to grow and increasing demands are placed upon already unstable water sources, the threat of conflict over water increases. India and Pakistan, for example, have experienced significant tensions over the use of the Indus River, as both rely upon its waters for hydroelectric power and irrigation. In southern Ethiopia, water and pasture-related conflicts between competing culture groups have led to hundreds of deaths and the forced departure of thousands of people from their lands (Waititu 2009a). Similar conflicts rage elsewhere on the African continent, as deforestation, climate change, and inefficient management have contributed to the dwindling of shared lakes and rivers. In the last half-century, total water withdrawals have tripled globally (Waititu 2009b). Droughts and increased use of as well as demand for water can cause desertification. According to the United Nations Convention to Combat Desertification (UNCCD), desertification is "land degradation in arid, semi-arid and dry sub-humid areas [drylands] resulting from various factors, including climatic variations and human activities" (UNCCD 1994).

Water shortage addressed in scholarly work is focused broadly, for example, on land use management (Caldwell 1984; Godwin et al. 1985), water scarcity (Curry 1985) and public policy (Durant 1985; Godwin et al. 1985; Hrezo and Hrezo 1985; Sander 1983). But since the 1980s, struggles over shared water resources have been growing and this has often resulted in conflicts among private corporations, the state, and citizens. Indeed, conflicts amongst people within communities at a local level have also arisen because of differential water needs, such as the needs of large versus small farms, which clearly point to a lack of homogeneous interests within villages, groups of villages, towns, or suburban areas (Birkenholtz 2009).[2]

The shifts in the ways in which water rights have been viewed, utilized, and managed across the decades can be viewed as transitioning across three stages: the First Water Age, the Second Water Age, and the Third Water Age (Gleick 2010). The First Water Age covered the period in which humans relied on the unpredictable national hydrological cycle to take what water was needed and to get rid of waste products. The transition to the Second Water Age began when humanity started to outgrow the limits of local water resources and to intentionally maneuver the hydrologic cycle, building dams, irrigation canals, and wastewater systems, and putting in place the first laws and social structures for managing water. This Age covered the nineteenth and twentieth centuries, when modern water systems were increasingly put in place that brought benefits to societies, but were eventually inadequate as billions of people still lacked safe water and sanitation. At the same time, aquatic ecosystems continued to be ravaged by overuse and were contaminated (Gleick 2010).

Conflicts over shared water resources are growing when the globe is transitioning to the Third Water Age as increasing efforts are being made to develop a sustainable system to use and manage this resource. It is in the Third Age (mainly the post-1980s) that water rights are couched within neoliberal thinking, intense protests and campaigns have emerged against the privatization of water, advocacy for community control of water (as a common property resource) has grown, and powerful campaigns against the bottling of water and bottled water have taken shape.

Thus, there has been a significant shift in scholarly work on water-related issues from the First and Second Ages to the Third Water Age. While much of the early attention was directed at mechanisms for harnessing water resources and in providing access to clean water primarily within the development studies literature, later work has focused on the struggles to control water as a public good and the privatization of water supplies.

EARLY APPROACHES TO THE STUDY OF WATER

Struggle over water wealth throughout history has been central to changes in societal habitats and the quality of ordinary lives. Solomon (2010) traces these historical struggles across civilizations and notes that water scarcity drives the political, economic, and environmental realities across the world. The special affinity between water and humans is reflected in water's primary role in certain stories in diverse cultures. Water was one of the Greeks' four primary terrestrial elements and one of five in ancient China and Mesopotamia. In fact, water continues to play a major role in common religious rituals (Christianity, Hinduism, Islam, Judaism, and Shinto). All of this reflects the value of this resource in everyday life.

At the height of the Second Water Age, scholarly work by social scientists was less concerned with water as a community right or central reason for resistance. The age of privatization had not yet taken hold and research was focused primarily on land use management (Caldwell 1984; Godwin et al. 1985), water scarcity (Curry 1985) and public policy (Durant and Holmes 1985; Godwin et al. 1985; Hrezo and Hrezo 1985; Sander 1983). Water was considered a public good that needed to be managed and the state was charged with monitoring access (Lovrich et al. 1985; Schneider 1982). As policies of deregulation gained popularity, the oversight of the state as a national monitoring body was loosened, ushering in the Third Water Age.

THE THIRD WATER AGE: THE CONTEMPORARY CRISIS

The new era of the Third Water Age, particularly under the pressure of scarcity, has led to a water market economy advocated by the state and enforced through private corporations, and is expanding the profit-seeking mechanism to colonize water. Chronic water scarcity undercuts the legitimacy of states and provokes social instability. Water riots, bombings, and many deaths occurred from 1995 to 2005 in various conflicts over water, for example, in Pakistan, India, provinces of arid North China, Cochabamba (Bolivia), between Kenyan tribes, among Somalia villages, and in the Darfur region of Sudan (Solomon 2010; Royte 2008).

In 1968, the ecologist Garrett Hardin wrote a paper called "The Tragedy of the Commons." This thesis about the rational pursuit of individual self-interest that leads to collective ruin fails to consider the complexity in collectively seeking control and ownership of the common good. Water is the ultimate commons. Seeming endless at one point in time, it has, as noted above, become increasingly scarce. In the face of a rapidly increasing global population, questions about the adequacy of the world's water supply challenge scholars and policy makers. Now Ecuador has become the first nation on earth to put the rights of nature in its constitution so that rivers and forests are not simple property, but maintain their own right to flourish. Under these laws, a citizen might file suit on behalf of an injured watershed, recognizing that its health is crucial to the common good (Kingsolver 2010).

Struggle over water wealth has been central to changes in societal habitats and the quality of ordinary lives throughout history. Control over and use of water, particularly when it is scarce, is certainly embedded in the local, national, social, and institutional arrangements that complicate the understanding of water rights. Agreeing to self-imposed limits instead, an unthinkable notion at first, will become the right thing to do. In addition, the collective aspect of rights to water is one important feature distinguishing it from many other types of property which can be individually owned and controlled.

In spite of the grim specter of water, individual activists and organizations are working to address the crisis. Many of them are reviving indigenous knowledge and techniques such as rainwater harvesting, along with developing norms and rules of use to manage resources as common property. At the center of these initiatives is resistance to the privatization of water resources, whether it is a river source or bottled water. This book is about the struggles of communities and citizens to resist the privatization of water. It is about community action to revive, maintain, and manage water resources; citizens' collective efforts to resist the privatization of public water resources, and campaigns against making water a commodity (such as by bottling).

As explained above, in the last half-century, total water withdrawals have tripled globally. In addition, the collective aspect of rights to water is one important feature distinguishing it from many other types of property which can be individually owned and controlled. This is the reason why struggles, spanning the local and global, over the use and management of water resources have intensified, but which also call for recognizing the gender, class, and caste dynamics in the processes of negotiating and resisting privatization trends.

Scholarship about access to and control of water has considered globalization and the power of the state in facilitating privatization. For instance, Shiva, in her book *Water Wars* (2002), reveals that conflicts disguised as ethnic wars or religious wars are conflicts over scarce resources such as water. This book celebrates the traditional role of water and calls for a movement such as the Cochabamba protests to retain rights to water. Cochabamba focuses specifically on Bolivia's privatization of water resources; it describes the solidarity of people in rejecting the transnational corporation that was taking over water resources. Later in 2005, Shiva provided an update on the struggles around the genetic engineering of food, the theft of culture, and the privatization of natural resources under the umbrella of the global resistance movement. The analysis draws attention to the principles of justice based on inclusion, nonviolence, and free sharing of the earth's resources. Similar topics are covered by Shiva in *Globalization's New Wars* (2005).

Concerns about access to quality water related to the privatization of urban water supplies has also been tackled by some scholars. For instance, Maude Barlow and Tony Clarke, two of the most active opponents of the privatization of water, emphasize inequalities in access to water in their book *Blue Gold* (2002). In doing so, they note that transnational corporations are increasingly controlling the world's dwindling water supply. In England and France, where water has already been privatized, rates have soared and water shortages have been severe. They also bring bottled water into the conversation because major bottled-water producers such as Perrier and Evian, among others, are buying up freshwater rights.

Urban water supplies are also the focus of the book *Privatizing Water* by Bakker (2010). Bakker, a geographer, addresses the question of how to supply an affordable, safe, and secure supply of water to the continuously expanding urban populations in underdeveloped countries. She covers a wide range of topics controversies regarding privatizing urban water systems, governance, and market failures that may affect water delivery, acceptance of water as a human right, the role of community in water supply management, and important hydrological and ecological factors that must be included in any attempt to solve water supply problems. However, a major portion of this book had appeared earlier as journal articles.

The urban basis of inequity to resources such as water is also at the center of the analysis in Gandy (2002), Kaika (2005), and Swyngedouw (2004). In contrast, Snitow and Kaufman with Fox (2007) focus solely on the United States to examine trends in the privatization of water. As bottled-water producers have bought rights to water, consumers have increasingly relied on bottled water. Both Gleick (2010) and Royte (2008) examine people's fixation over bottled water. In addition, scholars have examined water and markets (Simmons 2016; Bolens and Zwarteveen 2005), the role of gender and class (poverty) in access to water (Zwarteveen 1997; Brewster et al. 2006; UNWWAP 2015; Subramaniam 2014; Akiwumi 2015), and encouraging community participation in managing water resources (Pinto 2005; Prokopy 2005; Subramaniam 2014).

Although many of these works highlight globalization trends and the role of the state in organized resistance to privatization, analytical frameworks adopting a social movement lens have been limited.

Moreover, they do not capture local rural organizing of citizens who seek to set their own mechanisms for use of and access to water. The role of non-governmental organizations (NGOs) in rural organizing also requires analysis. In addition, globalization processes have increasingly facilitated the building of transnational coalitions to push for collective governance as well as challenging the privatization of water. The role of advocacy and transnational networks in resisting policies advocated by international institutions and global forums would be worth examining. For instance, how have transnational networks targeted international institutions such as the World Bank and national governments to resist the privatization of water? Consider, for instance, the role of the People's Water Forum (PWF)/Alternative World Water Forum (AWWF), a global network which represents the rural poor, environmental groups, and organized labor pushing for the elimination of water privatization. This forum has emerged in opposition to the World Water Forum (WWF), organized by the World Water Council (WWC) with support from major global financial institutions and national governments.

In order to assess the role of global institutional mechanisms, scholars have examined the role of international institutions such as the World Bank and the World Trade Organization (WTO) in promoting structural adjustment and trade liberalization, which has turned water into a prime battleground as regards the marketization and privatization of state-provided social services and public goods. Water marketization refers to the process of creating the economic and policy infrastructure for treating water as a marketed commodity. It refers to a broader set of linked transformations related to prices and property rights. Treating water as a scarce economic resource, a commodity, is a central tenet of the neoliberal agenda advocated by the World Bank.

All the above-mentioned conceptualizations—the state, globalization, neoliberalism, mobilizing, and collective action—will be discussed in Chapter 2. The central focus of this book is on local struggles and global networks that are interconnected and resist privatization, draw on local knowledge to develop innovative mechanisms to use and save water, and through the strategies adopted have drawn attention to their cause. The "ambivalent" state, I argue, controls ownership and rights to water through privatization policies created and enforced across the national and sub-national levels, and at the same time facilitates local communities to organize and use institutional mechanisms to demand control over water resources. My discussion of people's struggles for rights to water recognizes the differences in experiences based on the intersections of gender and class. Poor illiterate men and women may differ from educated people in terms of the strategies they adopt to demand ownership of water. Variations may also be evident in the knowledge about using and managing water resources. Such local protests are not entirely separate from the global networks engaged in articulating rights.

This Book, Cases, and Data

Many scholars working to transform knowledge in the area of social movements recognize that it is a contested terrain, since no single set of ideas or politics can explain the multiple ways in which people around the world have for decades mobilized, protested, organized, and constructed movements. More recently, and over the past 30 years or so, larger global forces that collectively come to be known as globalization have altered the conditions and patterns of lives of people around the world. The proliferation of multinational corporations, the creation of transnational economies, and their combined effect on local communities, the environment, and basic needs such as food and water, markets, and labor have become the most visible features of globalization.

In this book, I include a theoretical discussion of globalization and the state as comprising a complex network of institutions across the national and sub-national levels and "ambivalent" that is it advocates and adopts an uneven approach to neoliberalism. The primary tenet of neoliberalism that is of interest in this book is privatization. Local communities have targeted corporate and state power by responding to dispossession as privatization efforts have focused on accumulation for profit. The book will cover three cases to provide a local perspective on the mobilization and protests against privatization of water that will be tied in with a discussion of global forums such as the WWF and People's Water Forums/Alternative World Water Forum as broader transnational coalitions that support and/or challenge the privatization of water. In addition to the theoretical contributions of conceptualizing the state as "ambivalent," the analysis in this book has policy implications for both the state and international agencies that are invested in developing mechanisms for maintaining water supplies and ensuring access to clean water by including local communities, particularly marginalized groups, in dialogues.

This is also related to broader issues of environmental justice vital for consideration of intergenerational needs (Subramaniam and Zanotti 2015).

The three cases of local struggles discussed in this book are Tarun Bharat Sangh (TBS) and the Plachimada struggle, both in India, and the Concerned Citizens' Coalition of Stockton (CCCS) in California in the United States. TBS, an NGO led by Rajendra Singh, has been actively advocating for water rights in the state of Rajasthan. Let me provide readers with some basic information about Rajasthan, especially because water is scarce in this largely arid region. Rajasthan is a north-western state in India. It is the largest state in India by land area, covering an area of 342,239 square kilometers. The state's economy is primarily agricultural and pastoral. The north-western portion of Rajasthan is generally sandy, dry, and infertile, as most of this region is the Thar Desert, also known as the Great Indian Desert. Given its varied topography, the climate of Rajasthan greatly varies throughout the state. In the south-western part of the state, the land is wetter, hilly, and more fertile. Since rainfall in this state is quite limited, seasonal vegetation in the form of shrubs, grass species, and dwarf trees can only be found. In the plains, food crops are grown in areas where streamlets and rivers drain enough water for irrigation. Rajasthan is also a land of paradoxes. A famous tourist destination, the state attracts large numbers of domestic and foreign visitors due to its majestic forts, desert, colorful handicrafts, and vibrant folk culture. However, this colorful exterior veils traditions of the subjugation of women, child marriage, and seclusion that may limit their participation in the discussion about water resources, particularly in rural areas.

Singh, known as the "waterman," has become nationally and internationally known for protesting the privatization of water resources. In 1985, five young men of the organization, along with its secretary, Rajendra Singh, came to Kishori village in the Alwar district (Rajasthan). Grass was scarce for grazing cattle. Crop yields were low and only 3% of the cultivable area was irrigated. In such a grim situation, the TBS intervened as per the advice of the locals through the restoration of the traditional water harvesting structures, called johads (earthen check dams), in the region to capture and conserve rainwater, which improved percolation and groundwater recharge. In spite of opposition by government authorities, TBS continued to mobilize people to address water scarcity. Its greatest symbols are five rivers of the region, which have started flowing perennially after decades of drought, a direct result of conserving water in johads (Mahapatra 1999).

Government authorities opposed this organization as it directly interacted with the people and mobilized them for the work. But TBS continued to work with the mobilized community and organized them to consolidate the initiatives. By the end of the 1990s, TBS had expanded its work to 10 districts of Rajasthan (Menon et al. 2007). It is claimed that by 2005, TBS had worked with rural communities of 700 villages and had built thousands of water-harvesting structures in various parts of Rajasthan (Menon et al. 2007). TBS has been active in raising awareness about India's national water policy in rural areas of Rajasthan. In 2012, Rajendra Singh and the TBS launched "The Flow Research Network," which was intended to be a research network linking global communities to address community-initiated regeneration, traditional wisdom informing present-day action, rejuvenation of the balance between the man-made economy and the natural economy, a spiritual understanding of flow, and creating a bridge between the wisdom of the East and the West to re-create a global flow.

Unlike the TBS, the Plachimada struggle, in the southern Indian state of Kerala, comprises a loose network of local residents without a formal organizational structure. Kerala state has the highest literacy rate (almost 98%) in India and is a hotbed of a variety of labor/trade union movements, a literacy movement, and a movement of the fisherfolk, amongst many others. Plachimada is a small village in the Palakkad district of Kerala. It is situated in Chittur block's Perumatty Panchayat of the district.

In 1999, the Hindustan Coca-Cola Beverages Private Limited, a subsidiary of the Atlanta based Coca-Cola Company, established a plant in Plachimada, in the Palakkad district of Kerala. The Perumatty Village Panchayat (local village institution or council) gave a license to the company in 2000 to commence production of both aerated and nonaerated drinks, using glass and various forms of plastics for containers.[3] Coca-Cola drew around 510,000 liters of water each day from boreholes and open wells. For every 3.75 liters of water used by the plant, it produced one liter of product and a large amount of waste water.[4] Two years after production began, protests by local residents became commonplace. Local communities complained that water pollution and extreme water shortages were endangering their lives.

The next few years saw a confusing array of legal battles between the village council, the Kerala state government, and the Coca-Cola company. Coca-Cola offered to pay the *panchayat* President, who declined for fear of being co-opted. "Coca-Cola had created a water scarcity in a water abundant region" (Shiva n.d.: 2). The Palchimada struggle involved local protests and rallies, and local residents took legal recourse as well. This anti-Coca-Cola struggle is well-linked to the national network called the National Alliance of People's Movements (NAPM) and has attracted global attention. The Palchimada struggle and TBS differ from the third case, the CCCS, in that it is urban and based in the developed world.

In Stockton, California, a strong mayor played hardball in an effort to win the approval of the largest privatization contract in the West, while a grassroots coalition went to the ballot box and the courts to stop him. In 2003, Stockton, a city of 260,000 inhabitants, was experiencing an economic upswing with real estate development booming. Stockton is about 90 minutes north-east of San Francisco and is strategically located at the hub of California's complex hydrological system. The San Joaquin River flows through the city's downtown, and enormous pumps that provide water to southern California are located just south in the nearby town of Tracey. According to Snitow and Kaufman with Fox (2007), this puts Stockton at the pivot of the California water wars. The high real estate prices in the Bay Area make Stockton an attractive place for people to live.

The CCCS was formed in 2001 to monitor and challenge what it called the Stockton City Mayor Gary Podesto's push for privatization of the municipal water supplies. Podesto had a successful business career running grocery stores, which he sold to a major chain store and retired comfortably. With a lot of time on his hands and with experience in dealing with the government, he decided to run for the position of mayor of the city. He brought his business ideology into this elected position, the primary motive being profit. His business model, built on privatization, formed the basis for managing the city's water system.

Podesto argued that "a private company running the water department would not be privatization because the city would still own the water and the facilities" (Snitow and Kaufman with Fox 2007: 28). Members of the CCCS disagreed because such a partnership is privatization when a municipal function is turned over to the private sector. The major mission of the CCCS became the fight against the privatization of water. CCCS waged a grassroots campaign for four years that culminated in a legal victory to defeat the privatization of their municipal water utility. The company was required to return control of the utility to Stockton, which became effective in March 2008.

Such locally based organized or loose networks of challenges are not separate from the global networks of activists and organizations protesting privatization. Several water multinational corporations and international bodies such as the World Bank came together to create the WWC in 1996. The WWC organizes the WWF once every three years. The WWF engages in weeklong conferences about water usage, access, conservation, and sanitation. But this has not been straightforward. The WWC has direct links to two of the world's largest water corporations, Suez and Veolia.

As an alternative to the WWF, activists and scholars representing the rural poor, the environment, and organized labor organized the "People's Water Forum" (PWF, later renamed the "Alternative World Water Forum" or AWWF) to run concurrently with the WWF. The first PWF meeting was held in Florence in 2003 and attracted more than 1400 participants (70% of them Italian). There have also been Alternative Fora in 2005 in Geneva, in 2006 in Mexico City, in 2009 in Istanbul, in 2012 in Marseille, and in 2015 in Korea. Each of these Fora is organized by loose coalitions within countries or regionally based groups (such as East Asia, South America, Europe, and so on), involving a wide range of organizations, including transnational unions (for example, Public Services International, a global union of 20 million public workers in 150 countries, based in France), international environmental organizations and networks,[5] national and regionally based organizations,[6] and identity-based groups.[7] Many of these groups are formal organizations, but several are also loose informal networks or groups (cf. Aiyer 2007).

In an effort to organize the transnational water fora, key movement decisions are made through discussions or deliberations which are relatively accessible to some. However, because most of the Forums are in major cities across regions of the world, access to resources becomes key to being able to attend them. Representatives from various countries are therefore more likely to attend forums in their own geographical regions. Leaders of NGOs attend the fora and often affirm that they represent marginalized and rural communities (cf. Subramaniam 2014).

The recognition of the importance of access to a basic need such as water resonates in transnational forums. The rights frame provides a broad tent, enabling the many formal and informal organizations in the AWWF to join in the use of similar language (such as against the privatization of water), making the loose network act as a coalition. Nevertheless, activists have to overcome differences in perceptions and cultural differences, even differences in understanding what constitutes a right. I discuss these and other aspects of the global forums and the global water justice movement in detail in Chapter 4.

I use both primary and secondary data about each of the cases examined in this book. Primary data comprises interview notes, ethnographic notes, and public documents. I use primary data for the case of TBS in Rajasthan. I attended the 2007 December "water parliament" session of TBS which was attended by 190 village residents and about 30 guests. I use the observation notes from this session and informal conversations with TBS employees and villagers in the analysis. I also draw on information from a primary document, a 2006 Hindi publication of TBS which details the emergence of the water parliament as a local informal institution and deliberations in the parliament sessions from the first session in January 1999 up to the fifteenth session in December 2005. The publication provides an organizational perspective on the emergence and collective action of the villagers to resist privatization efforts of the state and corporations. I utilize secondary data in the cases of the Plachimada struggles and the Stockton protests. These sources include archival material accessed through the library and documents available on the internet. A very large volume of reports and other publicly available documents of the seven WWFs and six AWWFs form the basis of analysis of the global forums of water.

Overview of the Book

The unprecedented acceleration of transnational integration has created intersecting social, economic, technological, and political "global webs," which entangle and transform the lives of people across great geographic distances. The inescapable effects of globalization as well as the various local and national movements that have sought to transform the effects of globalization, such as the privatization of water resources, is central to this book. Because water is so critical for people, cultures worldwide share a belief that it belongs to everyone and that access to it is a basic human right (Gies 2009).

Chapter 2 will provide an overview of theoretical concerns relevant to understanding the rights to water that are intertwined between the local community challenges and the global-level transnational water forums. State institutions that form a complex web between the national and sub-national levels are the targets of challengers such as village-level residents or city-based citizens. In an era of globalization, the state is viewed by some as "retreating" and by others as "transforming," particularly in terms of adopting and implementing neoliberal tenets of privatization and accumulation. I integrate three main strands of scholarship—globalization and neoliberalism, the complexity of the state, and the struggles of mobilized people (village residents, urban citizens, transnationally)—to examine the ways in which people protest and negotiate for access to clean water as well as control over water resources. I argue that the 'ambivalent' state controls ownership and rights to water through privatization policies created and enforced across the national and sub-national levels, and at the same time facilitates local communities to organize and use institutional mechanisms to demand control over water resources. Local struggles are intertwined with the global networks that have sought to pressure global financial institutions and the state to reject neoliberal tenets that facilitate privatization.

Chapter 3 will examine three local struggles that demand rights to water by rejecting privatization. The economic dimension of globalization has involved specific attempts by the state (national governments) to privatize resources and assumes that markets will balance the supply, demand, and prices of resources that are viewed by people as a common property resource. This chapter sees water as a resource that has increasingly been privatized by the state, prompting community-based collective resistance to such policies. Community groups—village, town, and/or urban-based—have resisted environmental degradation and access as well as control over local sources of water. For instance, in the late 1990s, the TBS movement in Rajasthan engaged in the regeneration of the Arvari River and the rejuvenation of traditional systems for water harvesting. By holding a water 'parliament' twice a year to settle disputes and conflicts pertaining to use of water and pasture land, it has attempted to promote the democratic governance of scarce water resources. This collective initiative in a developing country like India is similar to the case of the CCCS in Stockton (California), which engaged in protesting the privatization of the city municipal water supply. In Stockton, a strong mayor proposed awarding a $600 million, 20-year contract to OMI-Thames, Inc.

to manage and operate its municipal wastewater utility. In response, the CCCS, which was formed in 2001, waged a grassroots campaign for four years that culminated in a legal victory to defeat the privatization of their municipal water utility. The company was required to return control of the utility to Stockton, which became effective March 2008. Although TBS is engaged in mobilizing people in rural areas and CCCS is urban-based, their goals are similar as they challenge privatization and demand a 'say' in who controls water resources. However, it is essential to acknowledge the role of TBS as an NGO which adds an additional dimension to the local–transnational nexus in challenges to water rights.

Unlike the CCCS and TBS, the Plachimada (Kerala state, India) struggle, the third case covered in this book, has involved the conservation of existing resources and protests against the excessive utilization of ground water by Coca-Cola. As they saw and experienced the depletion of their water resources, local residents came together in 2002 without any formal organization or leader. While Coca-Cola received support from the Kerala state government, the state High Court, local opposition political party leaders and the local panchayat (a local elected governance institution) responded in favor of the people's struggle. Adopting a legal route, the panchayat filed a case against the Kerala state. The Coca-Cola plant was shut down in 2005. The local panchayat as the local government institution pursued a policy that differed from that of the Kerala state government and the national government's agenda to attract and support private investment of all forms. This also shows that the multilayered structure of state institutions can enable citizens' protests. Yet local communities are not homogeneous and conflicts among groups over who determines rules of access to water also provide insights into negotiations within communities. I discuss and explain this in detail in Chapter 3. The analysis in this chapter seeks to unravel the notion of the "state," which is further problematized by the role of an NGO (in the case of TBS), the varying forms of protests, by residents, against dispossession. Gender and class will be considered in the analysis of the three cases.

The local struggles, irrespective of whether they are formally or informally organized, are connected to national activists and networks as well as global networks and forums. This aspect as related to global forums on water is the focus of Chapter 4. I will discuss the WWF and the PWF, (renamed the AWWF). Community-based collective rights to water have been raised, although with little success, in global forums such as the WWF. The WWF is organized by the WWC once every three years. The Council is an international think tank based in Marseilles, France, whose goal is to raise awareness of global water issues. To that end, the WWF plays host to political leaders, intergovernmental organizations, academics, NGOs and community organizations, UN agencies, and journalists from around the world. They engage in weeklong conferences about water usage, access, conservation, and sanitation. But this has not been straightforward. Critics of the WWF draw attention to the fact that several water multinational corporations and international bodies such as the World Bank came together to create the WWC in 1996. The WWC has direct links to two of the world's largest water corporations, Suez and Veolia. The Council promotes public–private partnerships that have put water services in some Latin American countries under private ownership. An alternative formation to the WWF is the AWWF, which represents the rural poor, the environment, and organized labor.

Activists have persuaded a block of Southern governments led by Uruguay to call for an alternative forum to be led by the United Nations. Their argument is founded on a desire among global water activists for the creation of a global framework of action that recognizes water as a right, thus eliminating the push for water privatization. I examine the role of the WWF and the AWWF based on three main themes. The first is their mission and goals and the organizations/groups involved in these forums as members and/or participants. The second theme concerns the analysis of how 'rights' to water are articulated by these forums, which will emphasize the similarities and differences in terms of how community rights to water (access and ownership) are incorporated (or not). Such an analysis will show how the WWF and the AWWF may differ in their perspectives to addressing the water crisis. The third theme I will examine is their policy prescription and on-the-ground actionable strategies.

All of these themes provide insights into the circulation of the discourse of rights to water between the local and the global in intertwined ways (not in any specific direction). Participants in the AWWF are mainly familiar with concerns in the location in which they organize people. They can speak to the issues in these specific communities in order to demand rights to control access to and use of water by targeting the state and/or private entities (small private organizations or large transnational companies). But they may not have a wider perspective on these issues and the members of the actual communities confronting the problems are not always present at these meetings. Rather, it may be the community organizers (who may or may not actually be from the communities in question) who attend the transnational meetings. The language—being able to articulate—used in transnational forums can also be a key factor in who represents and participates in transnational forums.

Chapter 5 will form the conclusion to the book. In this chapter, I discuss the theoretical contributions as well as the policy implications of the analysis. The analysis challenges the typical thinking that neoliberalism and specifically privatization is evenly adopted and implemented by the state. In reality, it is often rejected partially or wholly, which shows that the state is ambivalent. Such ambivalence is also partially attributable to the multilayered structure of state institutions. The analysis of the three cases will unravel the intricate links between state institutions, the community, and NGOs in struggles to seek access to water. In addition, a focus on the "community" involves examining the differences within locales, I will discuss how commonalities in concerns to seek rights to water can override differences of gender, class, and caste within communities.

Finally, the analysis of the contestation of rights to water shows how the micro-level (local) challenges and macro-level (global forums) discussions make interacting structures more visible and allow for the recognizing of power and politics as constitutive of water realities. In fact, the micro–macro connection will also provide a focus on the role of transnational corporations and the ways in which they exploit resources (ground water) of communities in the Global South.

Notes

1. See www.wri.org/our-work/topics/water (date accessed February 7, 2018).
2. I recognize the lack of a clear consensus in interests amongst people in using the term "community." By "community" groups, I mean local people's/citizens' groups or grassroots groups which are generally small and are formed in or across villages, towns, or cities working for a common cause, such as campaigning against the privatization of water. The complexities of focusing on the "community" are discussed in Chapter 2.
3. Panchayat is a local governance institution comprising elected members from the village.

4. These details are drawn from http://www.righttowater.info/rights-in-practice/legal-approach-case-studies/case-against-coca-cola-kerala-state-india/ accessed in July 2017.
5. For example, the 2012 meeting included Blue Planet, an international network of 100+organizations and thousands of individuals, based in California.
6. For example, the Council of Canadians, Forum italiano dei movimenti per l'acqua, Eau Secours! la Coalition québécoise pour une gestion responsable de l'eau, or the Comité Malien pour la défense de l'Eau.
7. For example, ACME Maroc is a group for Arabic people in France.

REFERENCES

Aiyer, Ananthakrishnan. 2007. "The Allure of the Transnational: Notes on Some Aspects of the Political Economy of Water in India." *Cultural Anthropology* 22(4): 640–658.
Akiwumi, Fenda A. 2015. "Analyzing Sierra Leone's Water Reform Efforts: Law, Environment, and Sociocultural Justice Issues." *Politics, Groups, and Identities* 3(4): 655–659.
Bakker, K. 2010. *Privatizing Water: Governance Failure and the World's Urban Water Crisis*. Ithaca, NY: Cornell University Press.
Barlow, Maude, and Tony Clarke. 2002. *Blue Gold: The Fight to Stop the Corporate Theft of the World's Water*. New York: New Press.
Birkenholtz, T. 2009. "Groundwater Governmentality: Hegemony and Technologies of Resistance in Rajasthan's (India) Groundwater Governance." *Geographical Journal* 175(3): 208–220.
Boelens, Rutgerd, and Margreet Zwarteveen. 2005. "Prices and Politics in Andean Water Reforms." *Development and Change* 36(4): 735–758.
Breakfast, Ntsikelelo, Gavin Bradshaw, and Richard Haines. 2014. "The Impact of Market-Friendly Policies in the Eastern Cape Municipalities." *Africa Insight* 44(3): 48–63.
Brewster, Marcia, Thora Martina Herrmann, Barbara Bleisch, and Rebecca Pearl. 2006. "A Gender Perspective on Water Resources and Sanitation." *Wagadu* 3: 1–23.
Caldwell, Lynton. 1984. "Land Use Policy as International Issue." *Policy Studies Journal* 12: 553–560.
Curry, Robert. 1985. "Mineral-Based Growth and Development-Generated Socioeconomic Problems in Botswana: Rural Inequality, Water Scarcity, Food Insecurity and Foreign Dependence Challenge New Governing Class." *American Journal of Economics and Sociology* 44(3): 319–336.
Durant, Robert, and Michelle Holmes. 1985. "Thou Shalt Not Covet Thy Neighbor's Water: The Rio Grande Basic Dispute and Its Implications for Interstate Groundwater Management." *Public Administration Review* 46(6): 821–831.

Gandy, M. 2002. *Concrete and Clay: Reworking Nature in New York City*. Cambridge, MA: MIT Press.
Gies, Erica. 2009. "Is Water a Human Right or a Commodity?" *World Watch*, March/April.
Gleick, Peter H. 2010. *Bottled and Sold: The Story Behind Our Obsession with Bottled Water*. Washington, DC: Island Press.
Godwin, R. Kenneth, Helen Ingram, and Dean Mann. 1985. "Symposium: Water Resources and Public Policy." *Policy Studies Review* 5(2): 349–352.
Groundwater Governance. n.d. Accessed February 16, 2017. http://www.groundwatergovernance.org.
Hrezo, Margaret, and William Hrezo. 1985. "From Antagonistic to Cooperative Federalism on Water Resources Development: A Model for Reconciling Federal, State and Local Programs, Policies and Planning." *American Journal of Economics and Sociology* 44(2): 199–214.
Johnson, Hope, Nigel South, and Reece Walters. 2016. "The Commodification and Exploitation of Fresh Water: Property, Human Rights and Green Criminology." *International Journal of Law, Crime and Justice* 44: 146–162.
Kaika, M. 2005. *City of Flows: Modernity, Nature and the City*. London: Routledge.
Kingsolver, Barbara. 2010. "Fresh Water." *National Geographic*, April.
Loftis, Randy Lee. 2016. *Could What Happened in Flint Happen Anywhere?* Accessed October 20, 2016. http://news.nationalgeographic.com/2016/01/160125-flint-michigan-water-crisis-lead-poisoning/.
Lovrich Jr., Nicholas, John Pierce, Taketsugu Tsurutani, and Takematsu Abe. 1985. "Water Pollution Control in Democratic Societies: A Cross-National Analysis of Sources of Public Beliefs in Japan and the US." *Policy Studies Review* 5: 431–450.
Mahapatra, R. 1999. "Waters of Life." *Down to Earth*, March 15.
Menon, Ajit, Praveen Singh, Esha Shah, Sharachchandra Lélé, Suhas Paranjape, and K. J. Joy. 2007. "Community-Based Natural Resource Management in Gopalpura, Rajasthan." In *Community-Based Natural Resource Management: Issues and Cases from South Asia*. New Delhi: Sage.
Montaigne, Fen. 2002. "Water Pressure." *National Geographic*, September.
Moyo, Khulekani, and Sandra Liebenberg. 2015. "The Privatization of Water Services: The Quest for Enhanced Human Rights Accountability." *Human Rights Quarterly* 37(3): 691–727.
Mullin, Megan. 2014. "Contested Water: The Struggle Against Water Privatization in the United States and Canada." *Review of Policy Research* 31(2): 148–150.
Olivera, O. 2004. *Cochabamba! Water War in Bolivia*. Cambridge, MA: South End Press.
Olivera, Marcela. 2014. "Water Beyond the State." *NACLA Report on the Americas* 47(3): 64–68.
Pinto, Jacintha. 2005. "Tank Water as Livelihood Support." *Integral Liberation* 9(2): 134–140.

Prokopy, Linda Stalker. 2005. "The Relationship Between Participation and Project Outcomes: Evidence from Rural Water Supply Projects in India." *World Development* 33(11): 1801–1819.

Royte, Elizabeth. 2008. *Bottlemania: Big Business, Local Springs, and the Battle over America's Drinking Water.* New York, NY: Bloomsbury.

Sander, William. 1983. "Federal Water Resources Policy and Decision-Making: Their Formulation is Essentially a Political Process Conditioned by Government Structure and Needs." *American Journal of Economics and Sociology* 43(1): 1–12.

Schneider, Mark. 1982. "Criminal Enforcement of Federal Water Pollution Laws in an Era of Deregulation." *Journal of Criminal Law and Criminology* 73(2): 642–674.

Shiva, Vandana. 2002. *Water Wars: Privatization, Pollution, and Profit.* Cambridge, MA: South End Press.

———. 2005. *Globalization's New Wars: Seed, Water, & Life Forms.* New Delhi: Women Unlimited, an Associate of Kali for Women.

Simmons, Erica S. 2016. "Market Reforms and Water Wars." *World Politics* 68(1): 37–73.

Snitow, A., and Kaufman D. with Fox M. 2007. *Thirst: Fighting the Corporate Theft of Our Water.* San Francisco: Jossey-Bass.

Solomon, Steven. 2010. *Water: The Epic Struggle for Wealth, Power, and Civilization.* New York: HarperCollins.

Subramaniam, Mangala. 2014. "Neoliberalism and Water Rights: Case of India." *Current Sociology* 62(3): 393–411.

Subramaniam, Mangala. 2016. "Politics of Rights to Water." *Blog, Gender & Society.* Accessed February 16, 2018. https://gendersociety.wordpress.com/2016/03/31/politics-of-rights-to-water.

Subramaniam, Mangala, and Laura Zanotti. 2015. "Introductory Essay: Environmental Justice-Just Livelihoods." *Politics, Groups, Identities* 3(4): 649–654.

Swyngedouw, E. 2004. *Social Power and the Urbanization of Water: Flows of Power.* Oxford: Oxford University Press.

UNCCD (United Nations Convention to Combat Desertification). 1994. *United Nations Convention to Combat Desertification in Those Countries Experiencing Serious Drought and/or Desertification Particularly in Africa: Text with Annexes.* Nairobi: UNEP.

UNESCO (United Nations Educational, Scientific and Cultural Organization). 2012. "World's Groundwater Resources are Suffering from Poor Governance." *UNESCO Natural Sciences Sector News.* Paris: UNESCO.

UNWWAP. 2015. *Water for a Sustainable World.* Paris: UNESCO.

Van der Gun, J. 2012. "Groundwater and Global Change: Trends, Opportunities and Challenges." *WWDR4 Side Publication Series No. 01.* Paris: UNESCO.

Waititu, Ernest. 2009a. "A Dwindling Existence for Africa's Pastoralists." *Frontline World: PBS.* Accessed February 16, 2018. http://www.pbs.org/frontlineworld/stories/africa705/history/africa.html.

———. 2009b. "Diminishing Water Resources Threaten Peace." *Washington Times*, May 26. Accessed December 21, 2009. www.pulitzercenter.org/open-item.cfm?id=1558.

World Water Assessment Programme. 2009. *The United Nations World Water Development Report 3: Water in a Changing World.* Paris: UNESCO; London: Earthscan.

Zwarteveen, Margreet Z. 1997. "Water: From Basic Need to Commodity: A Discussion on Gender and Water Rights in the Context of Irrigation." *World Development* 25(8): 1335–1349.

CHAPTER 2

Neoliberalism, the Ambivalent State, and Community Struggles

By asserting that flows of money and water follow universal, scientific laws and that human beings share the same aspirations and motives everywhere, neoliberalism establishes a universal rationality and efficiency, based on a "natural" and "objective" truth. The policies that are based on new institutional theories, in their turn, establish universal criteria for optimizing water management. Such universalization is a process of Foucauldian disciplining and normalization, while at the same time actively depoliticizing the water debate. Globalization processes and particularly the unevenness of neoliberalism, as discussed below, are key to understanding the rights to resources such as water.

Integrating three main strands of scholarship—globalization and neoliberalism, the state (as comprising several institutions), and the community that is involved in the struggles at the local level and the global level and is in that sense transnational (spanning across borders)—I posit that the state is at the center of the struggles against the privatization of water. By arguing that the state is a complex set of institutions across the national and sub-national levels which do not always work in tandem, the analysis can unravel the intricate links between the local and the global in the challenges created in seeking access to water. Acknowledging that a focus on the "community" requires examining the differences, by gender, class, and caste, within locales, I draw attention to how commonalities in concerns about the need for access to a basic need—water—can override differences. I rely on a social movement approach to examine the contestation of rights to water by including both micro-level (local) and macro-level discussions of struggles to make interacting structures more visible. An analysis of the struggles makes it possible to recognize power and politics as constitutive of water realities. Studying these processes of organizing and collective action will facilitate the bridging of the gap between forms of scholarship in two different social contexts—the developed country context (the United States) and the developing country context (India)—and will connect the local struggles and global forums about water rights.

The framework I develop will depart from some early work on collective action and management of resources which emphasize neoclassical and new institutionalist formulae which seemed attractive due to their clarity and the efficiency with which they simplify complex realities and behaviors. The neoinstitutional approach treats all organizing and organizations as formal in terms of structure and function. It overlooks the informal mechanisms of protest utilized by communities as well as the role of non-governmental organizations (NGOs) in representing communities in transnational forums. I begin with a discussion of globalization and the state.

GLOBALIZATION, NEOLIBERALISM, AND THE STATE

Globalization is a much-contested word. It is best understood as a spatial phenomenon, lying on a continuum with the "local" at one end and "the global" at the other. It denotes a shift in the spatial form of human organization and activity to transcontinental or interregional patterns of activity, interaction, and the exercise of power (Held 1998). Globalization, in its broadest definition, refers to the flow of capital, ideologies, people, media images, and technologies to different parts of the world (Appadurai 1996). Many scholars argue that the direction of these modernizing flows is from developed countries of the "North" to underdeveloped countries of the "South." Some of this literature describes such globalizing tendencies as benign and beneficial because of their modernizing tendencies. Others have viewed globalization from a longer historical perspective and attribute the flow of ideas, technologies, finances, and people from different centers—sometimes from the "South," sometimes from the "North"—and consider the influences to be both beneficial and problematic, depending on the nature of power differences (Sen 1999). In other words, flows are not unidirectional, but circuitous. Others, who have focused on the current phase almost exclusively, are highly critical and highlight the contradictory effects of globalization (Rai 2002).

The current phase of globalization contains traces of earlier exchanges and earlier economic and political structural adjustments. Globalization involves a stretching and deepening of social relations and institutions across space and time (Tabb 2004; Held 1998; Harvey 1990). However, "the global is everywhere ... Global is not 'out there' but transforms and redefines the meanings of domestic and international" (Tabb 2004: 41). Time is separated from space such that "place becomes increasingly *phantasmagoric*: that is to say, locales are thoroughly penetrated by and shaped in terms of social influences quite distant from them" (Giddens 1990: 19). Social relations and exchanges are being separated from territorial frameworks (Scholte 1997) presenting significant challenges to territorial-based governance institutions. While Giddens argues that globalization entails "the intensification of worldwide social relations which link distant localities in such a way that local happenings are shaped by events occurring many miles away and vice versa" (1990: 64), Held et al. (1999) suggest that globalization entails a transformation in the organization of social relations, generating networks of interaction and power across and within borders of the nation-state facilitated by the time-space compression (Harvey 1990; Giddens 1990).

In fact, the world is becoming more connected (McGrew 1998) and previously isolated social events are affecting different locales and populations with increasing consistency. Not only are locales and people affected by these processes, but national and regional economic systems and transactions are also influenced by easier and quicker flows of people and information across borders (Harvey 1990; Giddens 1990). These processes are important for considering the intertwined and circuitous connections of deliberations and struggles, across the local and global, for access to and control over resources such as water. Therefore, an examination of globalization also requires an analysis of transnational corporations, global financial institutions, global networks or forums that support or reject neoliberalism, and social movements that emphasize social justice. In addition, international development agencies, such as international and local NGOs, also contribute to the complexity of the local community-based politics in the developing world by providing development assistance or aid.

Although globalization in itself is not a new phenomenon, one of its contemporary features is the neoliberal agenda. As discussed below, locally neoliberalism may take the form of privatizing basic goods such as water, an act contributing to social inequality (Subramaniam 2014).

Scholars have examined globalization processes and the role of citizens negotiating and resisting the privatization of water resources that are largely perceived by the community as common property resources (Shiva 2002; Olivera 2004; Snitow and Kaufman with Fox 2007, among many others). Interestingly, much of this scholarship assumes a monolithic state and an even implementation of the neoliberal agenda, and pays less attention to the implications of the gender, caste, and class dynamics locally and in the global networks that have sought rights to water.

Neoliberalism

Neoliberalism is a theory of political economic practices that proposes that human well-being can be advanced by "liberating individual entrepreneurial freedoms and skills within an institutional framework characterized by strong private property rights, free markets, and free trade" (Harvey 2007: 2). It draws from the works of classic liberal economists such as Adam Smith, who in the late eighteenth century argued for the abolition of government intervention in economic matters. The ultimate (unreachable) goal of neoliberalism is a world where every action of every being is a market transaction, conducted in competition with every other being and influencing every other transaction, with transactions occurring in an infinitely short timeframe and repeated at an infinitely fast rate.

Globally, key institutions responsible for driving the neoliberal agenda are the Bretton Woods Institutions, which include the World Bank, the International Monetary Fund (IMF), and the World Trade Organization (WTO). These institutions, dominated by developed and globally powerful countries, advocate trade policies that disadvantage the poor in the developing world. These include providing heavy subsidies for production in rich countries, lowering trade barriers in developing nations for food commodity exports from developed countries, and pressuring poorer countries to export crops (Tabb 2004). Neoliberal policies created by the World Bank and IMF in the 1970s and 1980s removed tariffs on trade in order to "[allow] the unregulated market to determine the most efficient allocation of resources" (Green 2003: 12). These institutions restructure the economy to reduce the role of the state so that the private sector can assume a more prominent role. Thus, deregulation and the privatization of services are key aspects of neoliberalism packaged as structural adjustments and advocated to national governments.

The state, international institutions, and local people respond to neoliberal processes differently. Some national governments have adopted and implemented neoliberalism, a set of economic policies on the role of government and markets that have become widespread since the late 1970s. And some like Venezuela, Cuba, and Brazil, as well as a few countries in Africa, have recently resisted their adoption. A general characteristic of neoliberalism, advocated by the World Bank and the IMF, is the desire to intensify and expand the market by the formalization of transactions, making even a basic necessity such as water a commodity. Neoliberalism is therefore at the foundation of water marketization norms. Neoliberalism is a theory of political economic practices that has become widespread during the last 25 years or so.

The state is central to the implementation of the neoliberal agenda advocated by international institutions such as the World Bank, which itself is tied to the conditionalities that national governments are compelled to adopt in order to receive structural adjustment loans. The state enables the securing of private property rights and ensures by force, if necessary, the appropriate functioning of markets. Moreover, if markets do not exist for basic resources such as land and water, then they can be created by state action (Harvey 2003). The neoliberal order, supported by powerful states in the developed and the developing world, has been expanding and engaging wealthy corporate interests. At the same time, as discussed below, neoliberal policies and particularly privatization efforts, have been challenged by social movements, both locally and transnationally.

The most potent source of support for water marketization norms has been the broader phenomenon of neoliberal economic adjustment, grounded in a set of processes that have been working their way through the global economy since the late 1970s. Since 2000, the World Bank, rather than the IMF, has become a more significant international institutional actor advocating for more "streamlined" structural conditionality (Conca 2006). Along with the neoliberal adjustment, international trade rules were increasingly brought in place by the WTO (Rai 2008; Shiva 2002; Shiva et al. 2002). If the main effect of privatization pressures has been to weaken the state's role in allocating water, the main effect of trade liberalization has been to create a framework of rules that promote the mobilization of water as a marketable commodity. The state has been a major actor in directing privatization efforts, particularly corporate investment, as part of the neoliberal program. However, the broader effects are a contested process of struggles among national capitals, groups of citizens, the state and intergovernmental institutions (such as the World Bank and the IMF). This struggle has intensified as the neoliberal agenda has been advocated to national governments as central to the survival of economies in a global economy.

Scholarly studies of privatization of resources have drawn on David Harvey's well-known conceptualization of "accumulation by dispossession" and include a focus on water (cf. Bakker 2005; Swyngedouw 2005; Spronk and Webber 2007; Roberts 2008). "Nature itself has long resisted commodification," argues Swyngedouw (2005: 87), "but in recent years, nature and its waters have become an increasingly vital component in the relentless quest of capital for new sources of accumulation." Accumulation by dispossession refers specifically to practices that operate and evolve through the violent expulsion of peasants through land privatization, the conversion of common property rights, the suppression of alternative and indigenous forms of production and consumption, and the appropriation of natural resources (Harvey 2003).

Accumulation by dispossession makes available a set of assets at low or zero cost and "turn[s] them to profitable use" (Harvey 2003: 149). Profiting through privatization is a major precept of neoliberalism. Contributing to the discussion of neoliberalism, critical geographers assert that there is "neoliberalization" of socio-nature, a term that is used to highlight the particular ways in which specific "local neoliberalisms" are embedded in broader structures and relations of neoliberalism, which is heterogeneous and contested and which I utilize to analyze struggles for water rights (Bakker 2005: 544; also see Peck 2001; Peck and Tickell 2002).

Challenges to local neoliberalisms are complex as they involve politics of inequities as well as institutions, such as NGOs that may reinforce state interests. Such challenges can be explained using the concept of accumulation by dispossession. The local struggles shape and transform opportunities by advancing claims and challenging power within countries, and at the same time may contribute to broader anti-system movements. How is accumulation by dispossession articulated and practiced by the state? How have local citizens resisted the accumulation by dispossession that is also connected to networks across borders? And what are the challenges in mobilizing local residents to maintain water resources? I argue that the collective governance of water resources by citizens is contentious with the state at the center. In developing a framework for analysis, it is also important to problematize how we conceptualize and understand the state.

The Role of the State

Marginalized people's struggle for subsistence rights in the neoliberal era has theoretical implications for understanding the role of the state in a globalized world. Common to the three cases examined in this book is state institutions. The state, according to Kahler, is a "complex apparatus of centralized and institutionalized power that concentrates violence, establishes property rights, and regulates society within a given territory while being formally recognized as a state by international forums" (2002: 40). More importantly, states share some common characteristics, such as a system of laws and norms that regulate interactions among those who live in the state (Levi 2002) and thereby determines the ways in which resources are governed. Early conceptualizations of the state, such as that by Nettl (1968), have sparked various arguments on the role and nature of the state. Scholars suggest three conceptualizations of state responses that I refer to as representing retreat, centrism, and transformation.

State retreat scholars claim that economic globalization is eliminating the nation-state's macro-economic independence. For instance, Cable (1995) suggests that nation-states have lost economic sovereignty to both global institutions and financial markets as each has infiltrated economic processes once solely dominated by national governments. Following a similar thread, Strange (1996) argues that state authority is being diffused due to structural changes in the global economy. As states "compete" with other regulatory agencies, "states are less and less able to rely on the effectiveness of their directives" (Rosenau 2003: 227). These arguments suggest that states are losing their regulatory capacity to nonterritorial economic actors and processes. This presents economic globalization as a zero-sum process: nonterritorial processes or actors gain regulatory power while national governments lose their capabilities. The counterpoint to this argument suggests that states are promoting economic globalization and are thus the primary actors in the global system. It is to this argument that I turn next.

State centrist scholars argue that the nation-state is still the primary actor in the global economy and has effectively promoted greater global interconnectedness. According to Sorensen (2006), cross-border cooperation is state-driven and markets are controlled by states. Even more importantly, states are autonomous organizations that "may formulate and pursue goals that are not simply reflective of the demands and interests of social groups, classes, or society" (Skocpol 1985: 9). This state autonomy argument positions a national government separate from external forces which is somewhat complicated, especially because globalization processes, as discussed above, have on the one hand facilitated greater integration across borders, and on the other hand have also led states to assert their separate and independent status. Moreover, people's struggles (social movements) also require us to reconsider the autonomy of the state (for more on this aspect, see below). State centrist scholars argue that neoliberal policies promote growth (Bhagwati 2004) and national governments alone can promote industrialization (Kohli 2004). In the state-centrist position, economic globalization is still viewed as a zero-sum game, but the state is securing its position or consolidating power both within the country and in the global system.

The aforementioned arguments dichotomize state responses to economic globalization. However, this bifurcated system distracts attention from a more nuanced state argument. States are often more transformative, in that governments both initiate and respond to global externalities. As suggested by several scholars (cf. Weiss 2003; Hobson 2003; Cerny et al. 2005), states have the maneuverability to adapt to changing global circumstances with domestic policy responses. Sassen (1996) furthers the transformative argument by illustrating that state sovereignty is being transformed and decentered; global actors now must abide by a multiplicity of institutional arenas. State transformative arguments suggest that domestic political authorities have the capabilities to further entrench globalization processes while simultaneously responding to its outcomes. States should not be viewed as inactive global actors or the only players altering the system; instead, scholars should recognize the state's transformative capacity in the current global system.

The above-discussed conceptualizations of the state—retreat, centrism, and transformation—overlook two key aspects about the state. The first is the multi-institutional structure of the state and the second is the varying responses of these institutions to people's struggles and resistances. The state cannot be conceptualized as a monolithic structure or institution, but as "different arenas and sites of collective negotiation, coalition building, and struggle" (Bergman 2004: 219). These arenas could include the larger political apparatus of the nation that enacts policies as well as national and sub-national institutions that enforce and interpret them. State institutions—such as the judiciary and elected local governments—may also vary in terms of the roles that they perform.

Moreover, state-based political-economic decisions do not occur in a social vacuum. Evans (1997) suggests that a state-society synergy can prevent systemic economic shocks and thus protect the local citizenry. This argument neglects consideration of state actions that are antithetical to local community concerns and struggles. It also overlooks the various ways in which different state institutions, across the national and sub-national levels, may respond to community resistance and localized knowledge about resources, such as the case of water examined in this book. These aspects also have implications in terms of which neoliberal tenets are adopted and how they are implemented by the various state fractions, which are in turn further heightened by people's resistance. This results in uneven forms of neoliberalism being implemented by an ambivalent state.

THE AMBIVALENT STATE AND PEOPLE'S STRUGGLES

The role of the state is to create and preserve an institutional framework to maintain neoliberal practices through the validation of markets. It enables securing private property rights and ensuring, by force if necessary, the appropriate functioning of markets. Neoliberal policies in some form have been embraced by all states, including New Zealand, South Africa, and India. The neoliberal state should favor strong individual private property rights to facilitate economic development and the improvement of human welfare. The emphasis on private property rights is intended to serve as a mechanism against the so called "Tragedy of the Commons." The neoliberal framework puts serious limits on democratic governance. Yet it is important to note that states have not evenly embraced neoliberalism or, more specifically, privatized social services. In fact, globalization processes, as discussed above, have on the one hand facilitated greater integration across borders, and on the other hand have also led states to assert their separate and independent status in the global economy. People's struggles and resistance to neoliberal tenets such as privatization can also lead to the reversal of decisions by state institutions at the national and sub-national levels. Such inconsistencies and contradictions make for what I call an *ambivalent* state.

An ambivalent state pursues conflicting goals in adopting and implementing neoliberal tenets, which may vary (or not) across national and sub-national-level state institutions. As noted above, state structure comprises institutions at the local (village or city) level, the national level, and in between, that may not work in tandem with each other, making it possible for challengers to strategically target the various institutions to resist and reverse privatization (cf. Krishnan and Subramaniam 2015). Expressions of ambivalence include the flexible and variable decisions about privatization, between state institutions that may not comply with those advocated by global financial institutions such as the World Bank and the IMF. Some of these expressions have been prompted by people's action. For instance, in India, people's protests against privatization were related to being left at the mercy of the free-market system, with businesses and corporations hardly being held accountable for their lack of scruples. The ambivalent state's approach is not rational, but is dynamic as it combines competing aspects of neoliberalism based on the issue and shifts over time. Such inconsistencies are at the center of the uneven adoption and implementation of neoliberal tenets.

Neoliberalism is an uneven discourse across the global, national, and local levels. Local neoliberalism is atypical in that it brings within its fold localized institutions such as NGOs, engaged in mobilizing citizens and possibly contributing to accumulative practices. In addition, privatization—a major tenet of neoliberalism—is a relational concept and is not based exclusively on ownership. The relational notion compels us to consider issues of equity and efficiency and to recognize the local politics of inequalities that influence the supply of and access to water.

I suggest that the processes of commodification and privatization by the state serve as an impetus for mobilizing local residents to manage water resources. Such collective mobilizing may exacerbate inequalities based on gender, class, and caste, and may limit the possibilities of collective governance. The cases analyzed in this book show that accumulation by dispossession redefines the common property of water by appropriating a basic resource and suppressing local notions of consumption and maintenance. In essence, citizens have challenged the private corporation's quest for the accumulation of profits as the ambivalent state and the state apparatus do not always enforce the neoliberal goal.

Gender, Caste, and Class in Local and Global Alliances Across Borders

The nation-state may be implicated in the process of marketization and globalization, but it is constantly being shaped by the multiplicities of challenges and struggles of community groups. Community-led efforts to gain ownership of water resources, as I will discuss below, are complicated by the effects of the intersections of gender, caste, and class. Gender—the socially constructed, relational, structural, and symbolic differences between men and women—is a fundamental structuring mechanism in contemporary societies (Ferree et al. 2000). It is a primary way of signifying relations of power (Scott 1986) and of constructing privileges and hierarchies. Gender is not only about women; it refers to a structural relationship between sex categories which is linked to the state, the economy, and to other macro-level and micro-level processes.

In different ways and for a variety of reasons, all cultures use gender as a primary category of social relations. As a social process, gender intersects with race and class. Thus, gender is manifested in different ways, depending on one's position in the race and class system. Sociological analyses of gender emphasize that gender, like race and class, is a social experience of all. Feminist scholars have persuasively argued for analyses that see race, class, and gender as intersecting and interlocking systems of oppression in countries such as the United States (Hill-Collins 1990; Andersen 1993). This model can be usefully applied to the inequities of gender, caste, and class in India (Subramaniam 2006, 2012).[1] Although the origins of the caste system are widely debated, it is widely acknowledged that caste is a system of social stratification based on birth. At the highest level are the Brahmins (the priestly class) and at the lowest level are the Shudras (the untouchables and other lower castes). In between are the Kshatriyas (the warrior caste) and the Vaishyas (those engaged in business and commerce) (Subramaniam 2006). Cultural norms of seclusion and segregation can exacerbate gender inequality by their assumptions about what men and women need and what they are entitled to.

Women challenge gender (and class and caste)-based power through daily acts of individual or collective resistance (Scott 1985) and even in seeking access to and control over water resources. Acts of resistance enable disadvantaged individuals to negotiate the circumstances of their lives and gain power (Bell 2001). A significant body of scholarship particularly draws attention to women's localized protests against the effects of globalization (cf. Basu 1995; Purkayastha and Subramaniam 2004; Naples and Desai 2002; and Desai 2016, among many others). More importantly, such resistance can be organized through community groups.

Community-based groups are informal groups that often involve mobilization by an external facilitating organization (Subramaniam 2006) and are referred to as *grassroots* groups. Scholars have pointed out the limits to the term "grassroots" and, in particular, who should be considered as grassroots and whether this should include marginalized groups or coalitions among others (cf. Mahler 1998). In addition, studies draw attention to the negative use of the term "grassroots." For example, white and middle-class women used the term "grassroots" to refer to black and rural women in discussions of interventions by NGOs in South Africa (Mindry 2001). However, as noted by Naples: "This construction of the 'grassroots' fails to capture the politics of accountability and the extent to which so-called grassroots groups are inclusive and encourage participatory democratic practices" (2002: 7–8).

Grassroots groups, which are sometimes known as base groups, people's organizations, or local organizations (Bystydzienski and Sekhon 1999), emerge and/or work at the local level to improve and develop their communities either through community-wide or more specific memberships, such as women or farmers. Grassroots groups are community-based, generally small in scope and scale, and often focus on issues that directly impact members' lives. They often perform a mediating role between the private and the public, and between the state, the local community, and the family. The context within which grassroots groups emerge, such as the life experiences of members and/or program goals, can also shape group structure. Older groups may develop the potential to influence and even alter state policies and structures at the local community level.

Grassroots groups are neither bureaucratically structured nor entirely collectivist in form because of the ways in which they may engage in power sharing and distribution of authority. They are similar to hybrid forms of organizations. Feminist scholars have persuasively argued for considering informal and hybrid forms of organizations (Purkayastha and Subramaniam 2004; Ferree and Martin 1995). In doing so, they highlight the importance of the process that is the "means" as well as the end for change (Ferree and Martin 1995). In a similar vein, Martin (1990) calls for examining the concrete forms and practices of feminist organizations and their transformative impact on members. Such transformative impact is possible in feminist grassroots organizations (cf. Subramaniam 2006, 2012).

Feminist organizations actively promote the mobilization of grassroots women in defining their own needs and speaking out for their own cause. Such empowerment implies not only engagement in the formal organizations of politics, but also in confronting institutional power expressed in terms of family, religion, and other social institutions.

In order to challenge paternalistic assumptions, women's movements need specifically to empower the most marginalized women. For example, in rural India, efforts have been made to organize women's groups or collectives at the village level, which serve as institutions that women call their own (Subramaniam 2006, 2012).

Grassroots organizing creates and expands spaces to enable a process that redefines the form and content of politics (Bystydzienski and Sekhon 1999; Purkayastha and Subramaniam 2004). For instance, grassroots groups may aim to gain greater representation for the disadvantaged, such as women, in local political institutions. The groups provide a space for women to articulate their experiences, listen to others, and consider individual and collective challenges to injustices. In fact, such groups are particularly effective in consciousness-raising and promoting group solidarity through participation. Participation in such groups can be critical for the deeply disadvantaged, such as poor women, who have accepted silence and repression as part of their lives (see, for instance, the cases in Naples and Desai 2002; Subramaniam 2012).

Participation in locally based groups has been explored in the literature, primarily in reference to community organizing in the United States as well as in developing countries. Examples include black women during the Civil Rights Movement (Barnett 1993), community-based ethnic organization in the United States (Gilkes 1994; Chow 1987), and work in low-income communities (Naples 1998). This scholarship, as well as that on women's participation in local groups in developing countries (cf. Rose 1992; Tripp 2000; Subramaniam 2012), suggests that there are advantages to involvement in grassroots groups. However, discussions of participation often conflate "active" membership with "passive" followership in the form of attendance measures that do not necessarily capture the degree of engagement in a group. Participation involves interest and commitment to the group. It is about involvement in group activities such as discussion and decision-making, and speaking up in the group (Subramaniam 2012; Subramaniam and Zanotti 2015). People's participation in groups may somewhat be facilitated by NGOs.

Typically, development interventions in the "Third World" view women as beneficiaries. But in many developing countries, NGOs have become the vehicle for social development (Watkins et al. 2012), particularly as the state has partially or completely withdrawn services (such as education and healthcare), especially in rural areas of countries. The state has simultaneously facilitated the creation of NGOs to maintain access to international aid and to oversee development programs. Such an initiative addresses the criticism of the "top-down" clientalistic approach often adopted by the state, but fails to directly enable grassroots organizing. With new trends in multilevel power sharing, local democracy will increasingly have to be reconsidered in terms of a strictly delineated political sphere to one of social relations in a global civil society.

Localized grassroots groups were indirectly referenced by Frances Fox Piven in her 2007 presidential address to the American Sociological Association, "Can Power from Below Change the World," in which she argued for a new theory of "interdependent power." Drawing on the historical evidence in US history when independent revolutionaries, abolitionists, black power activists, and anti-Vietnam conflict protestors took their fights to the streets, Piven explained how political protests from those at the bottom of society with less access to traditional forms of power, the poor and marginalized, create change which can take the shape of political reforms. Taking on a litany of naysayers, Piven contended that this form of popular power has garnered the successes that it has because of neoliberal globalization. She found that interdependent power has less to do with resources, but is created in the process of social life, when cooperative bonds are formed in communities and networks. And in an age of neoliberal globalization where centralization and specialization increase alongside the division of labor, "webs of cooperation grow wider and more intricate, and the cooperative project involves more and diverse contributions from more and diverse people" (Piven 2008: 7). Thus, the potential for groups of people in the countryside to affect the decisions in the global cities, which are home to the headquarters of multinational corporations, increases to a new level. She contended that this power from below is a "force for change … [and] will ultimately determine whether another world is indeed possible" (Piven 2008: 1).

Scholars have increasingly been interested in the ways in which local grassroots groups connect with national and global groups to build ties and mobilize across borders (cf. Keck and Sikkink 1998; Moghadam 2005; Alvarez 2000; Bandy and Smith 2005; Naples and Desai 2002). Considering a wide variety of issues, including the adverse consequences of globalization, some of this scholarship draws on either a supranational approach or relies on the local to examine ties across borders. The strand of scholarly work that draws on a supranational approach focuses on analysis of international conferences, forums, and protests such as in Seattle (cf. Smith 2004, 2008; Rupp 1998; Keck and Sikkink 1998; Smith and Johnston 2002; Tarrow 2005). In contrast, work emphasizing the need for recognizing the locality empirically relies on data/information from only a single level to examine ties across borders (Subramaniam et al. 2003; Guidry et al. 2000; Moghadam 2005).

Scholarly work in the two strands mentioned above refers to movements across borders as global, international, or transnational. The terms "global" and "international" convey a sense of supranational action overlooking the significant localized efforts of actors which is somewhat captured in the term "transnational." The alternative conceptualization emphasizes local-level protests and action by organizations as integral components of transnational forms of resistance (Moghadam 2005; Piper and Uhlin 2004; Subramaniam et al. 2003). I build on scholarship that calls for considering transnational ties that contribute to the creation of loose informal networks (cf. Bob 2005; Anner and Evans 2004; Widener 2011).

Networks serve as a major enduring resource for transnational interaction (Keck and Sikkink 1998; Moghadam 2005; Smith 2001). Decentralized and nonhierarchical in form, networks are recognized as key to mobilization and have become institutionalized globally as an appropriate and effective means for movement activists to work together, despite differences in nationality, ethnicity, and class. Networks facilitate "people (or communities) to interact, to exchange information, to build social capital, and to mobilize for change. They help overcome distances that otherwise might appear insurmountable. And in so doing, they provide the basis for building movements" (Yashar 2005: 73). Movements across borders often involve transnational networks focusing on issues which are relevant across countries and involve mobilization across locations (Bandy and Smith 2005; Piper and Uhlin 2004).

In reality, many localized struggles are "internetworked social movements" (Langman 2005; Williford and Subramaniam 2015) that enable individual and group ties not only to electronically create virtual public spheres (Castells 1997), but also through alliances amongst leaders and citizens who have little in the way of organizational structure that can strengthen the demands of challenges across cities, towns, and villages. Some of these networks may be facilitated by NGOs that are more likely to receive funding from foundations or governments (Langman 2005). Such networks of challengers may extend beyond countries in the Global North and Global South, and, despite the differential power across these regions, can be powerful for targeting state institutions and resisting privatization advocated in global forums such as that for water (see below and Chapter 4). Internetworked movements, both global and regional, are characterized by circuitous flows of information and support for strategies that challenge neoliberalism and privatization of resources such as water.

Contesting Water Rights

Increasing water scarcity has been accompanied by an increasing push to treat water as an economic good and priced in a way as to recover the full costs of production directly from its users in some parts of the world (Budds and McGranahan 2003).[2] Making users pay for water and privatizing water is viewed as an effective and efficient means to address the increasing scarcity of water. Such privatization efforts, which are occurring in both developed and developing countries, are increasingly being resisted by citizens (cf. Bakker 2010). The April 2000 "water war" in Cochabamba, Bolivia, which some scholars characterize as a successful resistance to privatization, is of great interest to researchers (cf. Olivera 2004; Spronk and Webber 2007). Protests have also erupted in Argentina, Uruguay, South Africa, India, and other countries (Derman 2003; Bond 2005; *Wagadu* Special Issue 2006; Snitow and Kaufman with Fox 2007; Castro 2008; Shiva 2008; Royte 2008; Solomon 2010; Subramaniam and Williford 2012).

The UN Committee on Economic, Social and Cultural Rights has argued that treating water primarily as an economic good can undermine the realization of its value as a social and cultural good.[3] Moreover, the water as a commodity argument contradicts the common property notion of water or even the right to water as being a basic human right, which is one way in which locally based groups across countries view it. The common property notion distinguishes water from many other types of property "which can be individually owned and controlled" (Zwarteveen 1997: 1339). Rights exemplify social relations between people and are embedded in local cultural organization, making it about authority and power. In their discussion of Andean water reforms, Boelens and Zwarteveen call for "an alternative 'water rights ontology' that understands locally existing norms and water control practices, and the power relations that inform and surround them, as deeply constitutive of water rights" (2005: 735). So the most important question in relation to water is not whether to price, privatize, sell, or purchase, but rather who owns water and controls rights.

Rights to water encompass three dimensions: socio-legal, technical, and organizational (Boelens and Zwarteveen 2005). The socio-legal dimension is a right holder's claim to water and requires consideration of those who are excluded from its use. Such claims may vary based on state legislation and water laws, such as the 1981 Water Code of Chile

(Bauer 1997), or be based on local rules established and authorized by traditions and community organizations which are often contingent on social politics (Shiva 2002, 2005; Tarun Bharat Sangh 2006). The social politics could lead to some groups of local people having a greater say in the use and ownership rights (cf. Birkenholtz 2009; Olivera 2004). For instance, in Rajasthan, India decision-making rights and voting rights are in the hands of a powerful minority of large water right holders (Birkenholtz 2009). Generally, both legal and local reference water rights include specifications on how water control decisions are to be made and who can join this process. For instance, the main water users of the Central Valley of Cochabamba, Bolivia, are peasant and indigenous irrigator communities, which for decades have organized access to and distribution of water according to their "uses and customs" (Boelens and Zwarteveen 2005; Olivera 2004; Shiva 2002). For the first time in history, the rules governing water have been transformed by market forces. But the means (the technical dimension) to access water is essential, that is, from the source to the point of use, whether it is irrigation for farming or household needs.

Finally, it is necessary to organize and manage not just water allocation and the operation of infrastructure, but also the mobilization of resources and decision-making processes around these issues—the organizational dimension. And this too is evident in cases across the world. For instance, many irrigator organizations in the Andes are community-based, although some are set up or supported by governments or NGOs (Boelens and Zwarteveen 2005). In the Indian case of the Tarun Bharat Sangh (TBS) in the north-western state of Rajasthan, a parliamentary form of water governance has been adopted by the local village residents (Tarun Bharat Sangh 2006). As will be discussed in Chapter 3, the "water parliaments" (*pani sansad* in Hindi) are a forum for discussion about the use and control of water as a community resource. Striving for collective ownership of water is also directly linked to the participation of people for managing and using the resource, sometimes referred to as governance. By creating rules and limits on water use, collective water management may ensure sustainability and equity. Through the water parliament system, TBS has stressed self-governance through community participation. Both ideas—participation and community—need further examination.

In Peru, decision-making rights and voting rights are in the hands of a powerful minority of large water right holders; this has also been the tendency in Rajasthan, India generally, both legal and local reference to water rights include specifications on how water control decisions are to be made and who is allowed to join this process. For instance, the main water users of the Central Valley of Cochabamba, Bolivia are peasant and indigenous irrigator communities, which for decades have organized access to and distribution of water according to their "uses and customs" (Boelens and Zwarteveen 2005; Olivera 2004; Shiva 2002). For the first time in history, the rules governing water have been transformed by market forces.

Since the wave of water reforms beginning in the 1980s, private ownership of water utilities has increased markedly outside of the United States and Canada, where adoption has been slow (Whiteley et al. 2008). Two French international firms dominate the private sector and have interests in water projects in more than 120 countries (Gleick and Burns 2002). Private ownership follows different models, many of which are public–private partnerships. Past experiences suggest that, whether public or private, utilities work best when a strong, accountable municipal government maintains oversight over them. The case study of water utility privatization in Cochabamba, Bolivia by Baer (2008) illustrates that citizens expect to be able to participate in transparent and open decision-making processes in relation to their water service. Furthermore, in order to be acceptable, utility decisions must have consequences that people perceive to be equitable.

The privatized management of water utilities is a huge global sector, as is waste water services, dominated by corporate multinationals. In total, water is a fast-growing, highly fragmented, competitive, $400 billion-a-year industry (Solomon 2010). Subjecting water to the marketization and privatization forces is explosive, considering the high-profile conflicts across countries in which international corporations have been compelled to close or make costly modifications to their operations. Arguing that governments provide people with access to "adequate amounts of clean water and sanitation as a basic human right," Brewster et al. note that "large profits for private companies and denying access to poor people who cannot afford to pay are inappropriate" (2006: 13).

In Cochabamba, Bolivia, the citizens carried out a full-scale insurrection against privatization of their water by a global consortium led by Bechtel (Olivera 2004; Morberg 2005). Scores of privatization battles are taking place in the United States today. The case of Stockton discussed in this book is one example of a citizens' campaign against the city council's decision to privatize the municipal wastewater utility. As I will discuss in Chapter 3, after four long years, the coalition of citizens achieved a legal victory by preventing the privatization of their municipal water utility. Similarly, in India, the Plachimada struggle involved broad alliances and multiple strategies to demand the closure of the Coca-Cola plant in the southern Indian state of Kerala. The details of this struggle (Plachimada), which made written declarations that water is a common property resource, are discussed in Chapter 3.

The privatization of water resources is also at the heart of the increasing trend towards bottled water. In spite of resistance, private corporations have been in a scramble to commercialize water like other commodities such as oil and wheat. Bottled water is by far the world's fastest-growing beverage market, with global sales of $100 billion increasing at 10% per year and reshaping profits for corporate giants like Coca-Cola and Pepsi-Cola (Gleick 2010; Solomon 2010). Several international organizations have been proactive in organizing transnationally to challenge and protest the privatization of water. Maude Barlow of the Council of Canadians and Tony Clarke of the Polaris Institute in Ottawa are pioneers in the fight for water as a human right. Their 2002 book, *Blue Gold*, was a landmark in the international campaign against water privatization.

Campaigns and protests against the bottling of water and bottled water have gained ground across the globe. There is increasing and strong denouncement of transnational corporations like the Coca-Cola Company and PepsiCo, and associated but targeted attacks on corporate globalization expressing concern about the growing transnational corporate control of the planet's water resources (cf. Barlow and Clarke 2002; Bunker and Ciccantell 2005; Shiva 2002, 2005). The Cola Quit Plachimada movement (India) is similar to those occurring worldwide, particularly in peripheral nations like Bolivia, Uruguay, Argentina, and Ghana, which provide local resistance against the commoditization of water and against "corporate theft" (Barlow and Clarke 2002) and other forms of "water terrorism" (see Swyngedouw 2004).

Bottled water has become the mainstay of the Western world, particularly in the United States, mainly because of a fear of contamination of tap water (Gleick 2010; Royte 2008). It is also the source that people in some developing countries like India rely on. Water fountains that served as public sources of drinking water have been slowly disappearing as public water is increasingly pushed out in favor of private control and profit (see Gleick (2010) for a history of the water fountain). Confusion over water quality and contamination only furthers this trend. US Federal agencies given oversight over drinking water have no authority over bottled water, but the Food and Drug Administration regulate bottled water because it is considered a food product (Gleick 2010). *Corporate Accountability International*, an advocacy group that campaigns against bottled water, notes that bottled water is one of the least-regulated industries in the United States, much less than tap water. Moreover, instances of bottled water contamination are rarely reported to the public. These issues are not unique to the United States; similar controversies have arisen in other countries such as India as well (News Feature 2007).

If the privatization of water resources has been challenged and resisted by people who demand ownership and right to water, as discussed above, it is also essential to consider participation—which people participate and how—in controlling and managing water resources.

Participation in Managing Water Resources

States have often exercised their power on behalf of narrow, special interests. Politically subsidized water to farmers, urban residents, and hydropower users threatens sustainability as well as equity. Water development projects have generally not served the interests of indigenous peoples. Further, politically driven water projects have encroached on the water rights reserved for native peoples, and, at least in the United States, the federal government has been a poor steward of trust obligations to Native Americans. Political power, or the ability to marshal support in authoritative venues such as the branches and levels of government and media, is often decisive. Efficiency and equity (discussed below) may be important rationales employed in the process of building and wielding political power. But there are many cases in which political power serves neither of these values, yet still prevails.

Two major concepts that have driven discussions about control over water from an economic perspective are equity and efficiency. This emphasis is attractive due to the clarity and for the efficiency with which they simplify complex realities and behaviors, but they do not allow for the recognition of power and politics as constitutive of water realities. In their discussion of equity, Whiteley et al. note the need for "fair, open, and transparent decision-making processes" by encouraging participation of all affected groups (2008: 21). In fact, information and time as resources are important to ensure meaningful participation (a process) that is not dominated by elites. Scholars such as Whiteley et al. (2008) and Boelens and Zwarteveen (2005) link equity with process (participation) in order to make it efficient.

At the same time, discussion of such collective ownership requires scholars to be attentive to the complexities of participation of people. Participation in public decision-making and collective activities has very obvious costs and benefits in terms of time and effort. These costs and benefits affect men and women differently. For the poorest people, the opportunity costs of such participation may be prohibitive. Many poor households survive by hiring out their own labor on a casual basis. If half a day spent at a meeting means giving up half a day's paid labor, such people are unlikely to participate (United Nations 2005). These notions of costs and benefits are also tied in with considerations about the control and management of water resources from an economic perspective, which gives central attention to questions of equity and efficiency. This emphasis simplifies complex realities related to social power and local politics about water realities.

Costs and benefits of participation in collective processes and the corresponding economic paradigms of equity and efficiency assume a sense of rationality or what Ferree refers to as "individual profit maximizing behavior" (1992: 31). These approaches overlook the contributions of social movement scholarship to considering motivations beyond individual rationality as involving relationships and emotions; a sense of collective identity linked possibly to a shared history of struggles to maintain livelihoods.

The diversity in local users' interests and respect for the value of water reflects the broader equity principle of solidarity, which includes recognizing that each person or group has a right to participate in open and inclusive decision-making processes so that their values and visions can be considered fairly (Priscoli et al. 2004). Gender, class, and ethnicity are three important axes of societal divides that work in complex ways to color and shape an individual's possibilities for participation (cf. Boelens and Zwarteveen 2005; Zwarteveen 1997; Pinto 2005). For instance, water actors, users, private owners, and the state in the Andes are unequal in terms of monetary power, making the ability to pay for water a prime allocation mechanism that favors private companies and business actors. This could marginalize the users' involvement in decision-making in relation to managing the resource collectively. Moreover, market prices rarely adequately capture water's marginal value or utility to indigenous communities and are therefore a poor language for expressing this value (Boelens and Zwarteveen 2005). Nor are water actors equal in terms of their ability to access and influence decision-making and wield power in platforms and meetings; there are differences in terms of skills, information and education, which are often deeply rooted in historical divides and intertwined with cultural beliefs and biases. Consider, for instance, women's involvement in water resources.

In most societies, women have primary responsibility for management of the water supply, sanitation, and health at the household level, so their lives are deeply affected by the availability of clean water. Women and girls spend several hours each day collecting water, risking exhaustion, attacks, and drowning from floods (cf. Brewster et al. 2006; *Wagadu* Special Issue 2006; Singh 2006). Despite their deep connection to water resources, women are not recognized as having rights to water and they are often left out of the decision-making process at both the local and national levels. For instance, in her discussion of the need for the inclusion of women in discussions of the irrigation agenda, Zwarteveen (1997) calls for the recognition of women as farmers and irrigators, which is empowering for women. Programs such as the *Jala Samvardhane Yojana Sangha* (JSYS) have emphasized women's participation in efforts to revive local water tanks through community user groups in Karnataka (India) (Pinto 2005).

Formal committees and associations rely on public forms of decision-making and the transparent confrontation of issues such as access, distribution and rationing (examples include Tarun Bharat Sangh and JSYS in India, and water user groups in the Andes). However, there is plenty of evidence that poor people, and particularly poor women, are so dependent on reciprocal relations for their livelihoods that they are unlikely to take part in such discussions on a free and open basis. For example, they are likely to depend on their wealthier neighbors to hire them as seasonal laborers, to help with food in times of scarcity, or to allow them access to resources. Such relations of patronage result in a reluctance to openly confront neighbors, even when decisions that are being made are patently inequitable, and this preference for conflict avoidance is often reinforced by cultural beliefs in the desirability of living in peace and harmony and the dire consequences that will be incurred by those who upset this equilibrium (United Nations 2005).

Increasing the numbers of women and improving the balance between women and men in water management committees are desirable aims, but the presence of women on committees should not automatically be assumed to ensure gender equality. Societal structures and norms mean that even where women participate, they may not feel able to speak publicly, in front of men, particularly if this also means opposing men's views and interests. Additionally, there may be gendered norms regarding speaking in public (United Nations 2005). But mobilizing different groups of people is also directly related to recognizing their knowledge of water.

The right to water, as mentioned in Chapter 1, as an environmental issue is also about social justice. Increased participation in environmental governance and international bodies that uphold different rights has provided a multitude of avenues from which local communities can demand environmental justice.[4] These international developments have placed the overwhelming majority of concerns on distributional justice of resources such as water (Banerjee 2014). Attention to recognitional justice has been more of a central focus of the academic literature and not translated into action on the ground. The scope of justice efforts include challenging dominant environmental orthodoxies that identify local communities as "bad actors" and characterize the landscapes they call home as degraded or uninhabited (Forsyth 2003; Robbins 2012). The discourse of environmental politics and construction of nature are also what is at stake, as these powerful narratives and imaginaries shape ideas about what landscapes "should look like" (Neumann 1998). Identity politics are called up in these contexts, sometimes offering a reductive understanding of different ethnic, racial, and marginalized communities and their relationship to water. Both, participation" and "community" involvement in decision-making are complex, especially because local power politics may adversely affect the inclusion of the voices of women and the poor, among others (Agrawal and Gibson 1999; Rocheleau 2008; Subramaniam 2012, 2014).

The complexities involved in considering participation amongst local citizens to manage and use water has not deterred them from organizing in order to protest against privatization, whether it is in villages or rural areas, towns, or cities across the globe. Overall, communities, advocacy groups, and activists across the globe have been challenging and resisting privatization. While some scholarly work (cf. Boelens and Zwarteveen 2005; Zwarteveen 1997; Raman 2010; Aiyer 2007) utilizes concepts parallel to those in social movements, they are only limited in terms of analytical frameworks. Moreover, the role of global networks in promoting privatization, such as the World Water Forum (WWF), and resisting privatization (such as the People Water Forum/Alternative WWF) has been underexplored in this work. In fact, rights to water have also been raised with little success in global forums such as the WWF. The broader global water justice movement has actively sought to define the right to water as a human right.

From the early 1990s, a decade marked by the global environmental awakening symbolized by the first Earth Summit in 1992 in Rio de Janeiro, a consensus began to emerge about the effects of the declining freshwater resources on long term economic growth. This consensus led to a comprehensive worldwide assessment of major ecosystems launched under US auspices as the Millennium Ecosystem Assessment and completed in 2005 with input from experts from around the world. The assessment singled out freshwater systems as being unsustainable even at current demand levels. These initiatives have no doubt alerted societies to water scarcity and have somewhat opened up dialogues to consider possibilities, particularly through global forums. But the structure of these forums and who can participate in such global forums may be determined by available resources (such as funds necessary to travel to these forums) as well as the ability to articulate issues in a language that can be understood by participants from around the world. I will discuss the global forums on water in Chapter 4.

CONCLUSION

In this chapter, I draw on the scholarly literature about globalization and neoliberalism and the state as comprising several institutions across the national and sub-national levels to posit that there are intricate links of community struggles across the local and global arenas. As discussed above, the community is not a homogeneous group of people, but varies, such as through the intersections of gender, class, and caste, within locales. But commonalities in concerns about the need for access to a basic need—water—can override differences. An analysis of the struggles will allow for the recognition of power and politics as constitutive of water realities. The ambivalent state is at the center of these local and global struggles.

In an era of unbridled assault by neoliberal policies on citizenship rights, the role of the state has been a topic of debate in globalization literature (Evans 1997; Jessop 2010), where some argue that the state is irrelevant and others argue that the state paves the way for neoliberal reforms (DuRand and Martinot 2012; Tabb 2004). Departing from the either/or debates about the relevance of state institutions, I argue that the state is ambivalent, as it varies in its approach to adopting and implementing neoliberal tenets across the national and sub-national levels.

Both the state (along with international institutions) and local people respond to globalization and neoliberal processes differently. While the state responds actively by adopting and implementing a neoliberal agenda, local communities have resisted such policies which redefine water from the earlier common property resource to a marketable commodity that is controlled by private owners who profit from these shifts. Along with the neoliberal agenda, international trade rules were increasingly put in place by the WTO (Conca 2006; Shiva 2002; Shiva et al. 2002). If the main effect of privatization pressures has been to weaken the role of the state in allocating water, the main effect of trade liberalization has been to create a framework of rules that promote the mobilization of water as a marketable commodity. However, the broader effects are a contested process of struggles among citizen groups, the state, private corporations, and international institutions (such as the World Bank and the IMF). I argue that the state's implementation of neoliberal policies is inconsistent and such ambivalence can facilitate the broader social justice goals of communities such as the right to control and manage water resources.

The struggles over water resources span borders and the local networks are tied to global networks. Global networks in the struggles for water include those that have supported the privatization of water and those that have actively advocated for the human right to water. Ties between activists and organizations across borders enable the inclusion of the anti-privatization agenda in the global discourse. These activities also have implications for environmental justice.

NOTES

1. Under Articles 341 and 342 of the Indian Constitution, certain castes, specified by public notification, have been deemed to be Scheduled Castes (SC). Among the lowest in the caste hierarchy, SCs are still among the poorest sections of Indian society. "Untouchability," along with rituals and ritual prohibitions, are an essential feature of the caste system. The practice of untouchability isolates the "shudras" (or dalits) from those belonging

to the upper castes. More than one-sixth of India's population (about 160 million people) is categorized as dalits, meaning "downtrodden" (Government of India 1999). Considered unclean and hence untouchable, the SCs are subject to various types of discrimination, ranging from physical avoidance to exclusion from Hindu temples. See Joshi (1986) and Jogdand (1995) for more information on this. "Untouchables" may not cross the line dividing their part of the village from that occupied by higher castes. They may not use the same wells for drawing water or drink from the same cups as used by others in tea stalls. Although abolished by law, untouchability as a practice remains an integral part of daily life in rural India. The descriptive terms used to refer to them are "untouchables," "Harijan" (a term coined by Gandhi), and "dalits." I use all these terms interchangeably.
2. I call for recognizing the difference in the general changes in considering water as an economic good versus neoliberalism. For instance, water law in the United States is and has been heavily contested, such as in the case of the Owen River diversions, showing the collaboration or even corruption of state with private corporations.
3. For instance, Ecuador's 2008 Constitution asserts that water is a fundamental human right. Indigenous peoples and students took to the streets to protest and influence the drafting of the Constitution to protect "Mother Earth." A protest attended by the second author in March 2008 included several signs and chants critical of the neoliberal agenda and in support of water rights.
4. See Schlosberg (2007) for an in-depth discussion of the multiple dimensions of environmental justice.

References

Agrawal, Arun, and Clark C. Gibson. 1999. "Enchantment and Disenchantment: The Role of Community in Natural Resource Conservation." *World Development* 27(4): 629–649.

Aiyer, Ananthakrishnan. 2007. "The Allure of the Transnational: Notes on Some Aspects of the Political Economy of Water in India." *Cultural Anthropology* 22(4): 640–658.

Alvarez, Sonia. 2000. "Translating the Global: Effects of Transnational Organizing on Local Feminist Discourses and Practices in Latin America." *Meridians: A Journal of Feminisms, Race, Transnationalism* 1(1): 29–67.

Andersen, Margaret L. 1993. *Thinking About Women*. Boston, MA: Allyn & Bacon.

Anner, M., and P. Evans. 2004. "Building Bridges Across a Double Divide: Alliances Between the US and Latin American Labour and NGOS." *Development in Practice* 14(1): 34–47.

Appadurai, Arjun. 1996. *Modernity at Large: Cultural Dimensions of Globalization*. Minneapolis: University of Minnesota Press.
Baer, Madeline. 2008. "The Global Water Crisis, Privatization, and the Bolivian Water War." In *Water, Place, and Equity*, edited by John M. Whiteley, Helen Ingram, and Richard Perry, pp. 195–224. Cambridge, MA: MIT Press.
Bakker, K. 2005. "Neoliberalizing Nature? Market Environmentalism in the Water Supply in England and Wales." *Annals of the Association of American Geographers* 95(3): 542–565.
———. 2010. *Privatizing Water: Governance Failure and the World's Urban Water Crisis*. Ithaca, NY: Cornell University Press.
Bandy, Joe, and Jackie Smith (eds.). 2005. *Coalitions Across Borders: Transnational Protest and the Neoliberal Order*. Lanham, MD: Rowman & Littlefield.
Banerjee, Damayanti. 2014. "Towards an Integrative Framework for Environmental Justice Research." *Society and Natural Resources* 27: 805–819.
Barlow, Maude, and Tony Clarke. 2002. *Blue Gold: The Fight to Stop the Corporate Theft of the World's Water*. New York: New Press.
Barnett, B. M. 1993. "Invisible Southern Black Women Leaders in the Civil Rights Movement: The Triple Constraints of Gender, Race, and Class." *Gender and Society* 7(2): 162–182.
Basu, A. 1995. "Introduction." In *The Challenge of Local Feminisms: Women's Movements in Global Perspective*, edited by Amrita Basu with the assistance of C. E. McGory, pp. 1–21. Boulder, CO: Westview Press.
Bauer, C. J. 1997. "Bringing Water Markets Down to Earth: The Political Economy of Water Rights in Chile, 1976–95." *World Development* 25(5): 639–656.
Bell, Beverly. 2001. *Walking on Fire: Haitian Women's Stories of Survival and Resistance*. Ithaca, NY: Cornell University Press.
Bergman, Solveig. 2004. "Collective Organizing and Claim Making on Child Care in Norden: Blurring the Boundaries Between the Inside and the Outside." *Social Politics* 11(2): 217–246.
Bhagwati, Jagdish. 2004. *In Defense of Globalization*. New York: Oxford University Press.
Birkenholtz, T. 2009. "Groundwater Governmentality: Hegemony and Technologies of Resistance in Rajasthan's (India) Groundwater Governance." *Geographical Journal* 175(3): 208–220.
Bob, C. 2005. *The Marketing of Rebellion: Insurgents, Media, and International Activism*. New York: Cambridge University Press.
Boelens, Rutgerd, and Margreet Zwarteveen. 2005. "Prices and Politics in Andean Water Reforms." *Development and Change* 36(4): 735–758.
Bond, P. 2005. "Globalisation/Commodification or Deglobalisation/Decommodification in Urban South Africa." *Policy Studies* 26(3–4): 337–358.
Brewster, Marcia, Thora Martina Herrmann, Barbara Bleisch, and Rebecca Pearl. 2006. "A Gender Perspective on Water Resources and Sanitation." *Wagadu* 3: 1–23.

Budds, J., and G. McGranahan. 2003. "Are the Debates on Water Privatization Missing the Point? Experiences from Africa, Asia and Latin America." *Environment and Urbanization* 15(2): 87–113.

Bunker, Stephen G., and Paul S. Ciccantell. 2005. *Globalization and the Race for Resources*. Baltimore, MD: Johns Hopkins University Press.

Bystydzienski, J. M., and J. Sekhon. 1999. "Introduction." In *Democratization and Women's Grassroots Movement*, edited by Jill M. Bystydzienski and Jothi Sekhon, pp. 1–21. Bloomington: Indiana University Press.

Cable, Vincent. 1995. "The Diminished Nation-State: A Study in the Loss of Economic Power." *Daedalus* 124(2): 23–53.

Castells, M. 1997. *The Power of Identity*. Malden, MA: Blackwell.

Castro, J. E. 2008. "Water Struggles, Citizenship and Governance in Latin America." *Development* 51: 72–76.

Cerny, Philip G., Georg Menz, and Susanne Soederberg. 2005. "Different Roads to Globalization: Neoliberalism, the Competition State, and Politics in a More Open World." In *Internalizing Globalization: The Rise of Neoliberalism and the Decline of National Varieties of Capitalism*, edited by Susanne Soederberg, Georg Menz, and Philip C. Cerny, pp. 1–30. New York: Palgrave Macmillan.

Chow, Esther L. 1987. "The Development of Feminist Consciousness Among Asian American Women." *Gender and Society* 1(3): 284–299.

Conca, Ken. 2006. *Governing Water: Contentious Transnational Politics and Global Institution Building*. Cambridge, MA: MIT Press.

Derman, B. 2003. "Cultures of Development and Indigenous Knowledge: The Erosion of Traditional Boundaries." *Africa Today* 50(2): 67–85.

Desai, Manisha. 2016. *Gendered Geographies of Struggle Against Neoliberal Development*. New York: Routledge.

DuRand, Cliff, and Steve Martinot. 2012. *Recreating Democracy in a Globalized State*. Atlanta: Clarity Press.

Evans, Peter. 1997. "The Eclipse of the State? Reflections on Stateness in an Era of Globalization." *World Politics* 50(1): 62–87.

Ferree, Myra Marx. 1992. "The Political Context of Rationality: Rational Choice Theory and Resource Mobilization." In *Frontiers in Social Movement Theory*, edited by Aldon D. Morris and Carol McClurg Mueller, pp. 29–52. New Haven, CT: Yale University Press.

Ferree, Myra Marx, Judith Lorber, and Beth Hess (eds.). 2000. *Revisioning Gender*. Walnut Creek, CA: Altamira Press.

Ferree, M. M., and P. Y. Martin (eds.). 1995. *Feminist Organizations: Harvest of the New Women's Movement*. Philadelphia: Temple University Press.

Forsyth, T. 2003. *Critical Political Ecology: The Politics of Environmental Science*. New York: Routledge.

Giddens, Anthony. 1990. *The Consequences of Modernity*. Stanford, CA: Stanford University Press.

Gilkes, C. T. 1994. "'If It Wasn't for the Women ...': African American Women, Community Work, and Social Change." In *Women of Color in US Society*, edited by M. Baca Zinn and B. Thornton Dill, pp. 229–246. Philadelphia: Temple University Press.

Gleick, Peter H. 2010. *Bottled and Sold: The Story Behind Our Obsession with Bottled Water.* Washington, DC: Island Press.

Gleick, Peter H., and William C. G. Burns. 2002. *The World's Water, 2002–2003: The Biennial Report on Freshwater Resources.* Washington, DC: Island Press.

Green, Duncan. 2003. *Silent Revolution: The Rise and Crisis of Market Economics in Latin America*, 2nd ed. New York: Monthly Review

Guidry, John A., Michael D. Kennedy, and Mayer N. Zald. 2000. "Globalization and Social Movements." In *Globalizations and Social Movements: Culture, Power, and the Transnational Public Sphere*, edited by John A. Guidry, Michael D. Kennedy, and Mayer N. Zald, pp. 1–32. Ann Arbor: University of Michigan Press.

Harvey, David. 1990. *The Condition of Postmodernity.* Malden, MA: Blackwell.

———. 2003. *The New Imperialism.* Oxford: Oxford University Press.

———. 2007. *A Brief History of Neoliberalism.* Oxford: Oxford University Press.

Held, D. 1998. "Democracy and Globalization." In *Reimagining Political Community Studies in Cosmopolitan Democracy*, edited by D. Archibugi, D. Held, and M. Koehler, pp. 11–27. Cambridge: Polity Press.

Held, David, Anthony McGrew, David Goldblatt, and Jonathan Perraton. 1999. "Globalization." *Global Governance* 5(4): 483–496.

Hill-Collins, P. 1990. *Black Feminist Thought: Knowledge, Consciousness, and the Politics of Empowerment.* New York: Routledge.

Hobson, John M. 2003. "Disappearing Taxes or the 'Race to the Middle'? Fiscal Policy in the OECD." In *States in the Global Economy: Bringing Domestic Institutions Back in*, edited by Linda Weiss, pp. 37–57. New York: Cambridge University Press.

Jessop, Bob. 2010. "The 'Return' of the National State in the Current Crisis of the World Market." *Capital & Class* 34(1): 38–43.

Jogdand P. G. (ed.). 1995. *Dalit Women in India: Issues and Perspectives.* New Delhi: Gyan Publishing House.

Joshi, B. R. 1986. "Introduction." In *Untouchable! Voices of the Dalit Liberation Movement*, edited by B. R. Joshi, pp. 9–14. London: Zed Books.

Kahler, Miles. 2002. "The State of the State in World Politics." In *Political Science: The State of the Discipline*, edited by Ira Katznelson and Helen V. Milner, pp. 56–83. New York: W.W. Norton.

Keck, Margaret E., and Kathryn Sikkink. 1998. *Activists Beyond Borders: Advocacy Networks in International Politics.* Ithaca, NY: Cornell University Press.

Kohli, Atul. 2004. *State-Directed Development: Political Power and Industrialization in the Global Periphery.* New York: Cambridge University Press.

Krishnan, Preethi, and Mangala Subramaniam. 2015. "Understanding the State: Right to Food Campaign in India." *The Global South* 8(2): 101–118.
Langman, Lauren. 2005. "From Virtual Public Spheres to Global Justice: A Critical Theory of Internetworked Social Movements." *Sociological Theory* 23(1): 42–74.
Levi, Margaret. 2002. "The State of the Study of the State." In *Political Science: The State of the Discipline*, edited by Ira Katznelson and Helen V. Milner, pp. 33–55. New York: W.W. Norton.
Mahler, Sarah. 1998. "Theoretical and Empirical Contributions Towards a Research Agenda for Transnationalism." In *Transnationalism from Below*, edited by Michael Peter Smith and Luis Eduardo Guarnizo, pp. 64–101. New Brunswick, NJ: Transaction Publishers.
Martin, P. Y. 1990. "Rethinking Feminist Organizations." *Gender and Society* 4(2): 182–206.
McGrew, Anthony G. 1998. "The Globalisation Debate: Putting the Advanced Capitalist States in Its Place." *Global Society: Journal of Interdisciplinary International Relations* 12(3): 299–323.
Mindry, Deborah. 2001. "NGOs, 'Grassroots', and the Politics of Virtue." *Signs* 26(4): 1187–1211.
Moghadam, Valentine. 2005. *Globalizing Women: Transnational Feminist Networks*. Baltimore, MD: Johns Hopkins University Press.
Morberg, David. 2005. "Plunder and Profit." In *Beyond Borders: Thinking Critically About Global Issues*, edited by Paula S. Rothenberg, pp. 446–449. New York: Worth Publishers.
Naples, Nancy (ed.). 1998. *Community Activism and Feminist Politics: Organizing Across Race, Class, and Gender*. New York: Routledge.
———. 2002. "Changing the Terms: Community Activism, Globalization, and the Dilemmas of Transnational Feminist Praxis." In *Women's Activism and Globalization: Linking Local Struggles and Global Politics*, edited by Nancy Naples and Manisha Desai, pp. 3–14. New York: Routledge.
Naples, Nancy, and Manisha Desai (eds.). 2002. *Women's Activism and Globalization: Linking Local Struggles and Transnational Politics*. New York: Routledge.
Nettl, J. P. 1968. "The State as a Conceptual Variable." *World Politics* 20(4): 559–592.
Neumann, R. 1998. *Imposing Wilderness: Struggles over Livelihood and Nature Preservation in Africa*. Berkeley: University of California Press.
News Feature. 2007. "A Breath of Fresh Air." *Nature* 445(15): 706–708.
Olivera, O. 2004. *Cochabamba! Water War in Bolivia*. Cambridge, MA: South End Press.
Peck, J. 2001. "Neoliberalizing States." *Progress in Human Geography* 25(3): 445–455.
Peck, J., and A. Tickell. 2002. "Neoliberalizing Space." *Antipode* 34(3): 380–404.

Pinto, Jacintha. 2005. "Tank Water as Livelihood Support." *Integral Liberation* 9(2): 134–140.
Piper, Nicola, and A. Uhlin. 2004. "New Perspectives on Transnational Activism." In *Transnational Activism in Asia*. New York: Routledge.
Piven, Frances Fox. 2008. "Can Power from Below Change the World." *American Sociological Review* 73(1): 1–14.
Priscoli, J. D., J. C. I. Dooge, and R. Llamas. 2004. *Water and Ethics: Overview*. Paris: UNESCO.
Purkayastha, B., and M. Subramaniam (eds.). 2004. *The Power of Women's Informal Networks: Lessons in Social Change from South Asia and West Africa*. Lanham, MD: Lexington Press.
Rai, Shirin. 2002. *Gender and the Political Economy of Development: From Nationalism to Globalization*. Malden: Blackwell.
———. 2008. *The Gender Politics of Development: Essays in Hope and Despair*. New Delhi: Zed.
Raman, Ravi K. 2010. "Transverse Solidarity: Water, Power, and Resistance." *Review of Radical Political Economics* 42(2): 251–268.
Robbins, P. 2012. *Political Ecology: A Critical Introduction*. Malden, MA: Wiley.
Roberts, A. 2008. "Privatizing Social Reproduction: The Primitive Accumulation of Water in an Era of Neoliberalism." *Antipode* 40(4): 535–560.
Rocheleau, Dianne E. 2008. "Political Ecology in the Key of Policy: From Chains of Explanation to Webs of Relation." *Geoforum* 39: 716–727.
Rose, K. 1992. *Where Women Are Leaders: The SEWA Movement in India*. London: Zed Books.
Rosenau, James N. 2003. "Governance in a New Global Order." In *The Global Transformations Reader: An Introduction to the Globalization Debate*, 2nd ed., edited by David Held and Anthony McGrew, pp. 223–233. Cambridge: Polity Press.
Royte, Elizabeth. 2008. *Bottlemania: Big Business, Local Springs, and the Battle over America's Drinking Water*. New York, NY: Bloomsbury.
Rupp, Leila J. 1998. *Worlds of Women: The Making of an International Women's Movement*. Princeton, NJ: Princeton University Press.
Sassen, Saskia. 1996. *Losing Control? Sovereignty in the Age of Globalization*. New York: Columbia University Press.
Schlosberg, David. 2007. *Defining Environmental Justice: Theories, Movements, and Nature*. New York, NY: Oxford University Press.
Scholte, Jan Aart. 1997. "Global Capitalism and the State." *International Affairs* 73(3): 427–452.
Scott, James C. 1985. *Weapons of the Weak: Everyday Forms of Peasant Resistance*. New Haven: Yale University Press.
Scott, Joan. 1986. "Gender as a Useful Category of Historical Analysis." *American Historical Review* 91(5): 1053–1075.

Sen, Amartya. 1999. *Development as Freedom*. New York: Knopf.
Shiva, Vandana. 2002. *Water Wars: Privatization, Pollution, and Profit*. Cambridge, MA: South End Press.
———. 2005. *Globalization's New Wars: Seed, Water, & Life Forms*. New Delhi: Women Unlimited, an Associate of Kali for Women.
———. 2008. "From Water Crisis to Water Culture." *Cultural Studies* 22: 498–509.
Shiva, Vandana, Radha Holla Bhar, Afsar H. Jafri, and Kunwar Jalees. 2002. *Corporate Hijack of Water: How World Bank, IMF and GATS-WTO Rules Are Forcing Water Privatization*. New Delhi: Navdanya.
Singh, Nandita. 2006. "The Changing Role of Women in Water Management: Myths and Realities." *Wagadu* 3: 94–113.
Skocpol, Theda. 1985. "Bringing the State Back in: Strategies of Analysis in Current Research." In *Bringing the State Back in*, edited by Peter B. Evans, Dietrich Rueschemeyer, and Theda Skocpol, pp. 3–37. New York: Cambridge University Press.
Smith, J. 2001. "Globalizing Resistance: The Battle of Seattle and the Future of Social Movements." *Mobilization* 6(1): 1–19.
Smith, Jackie. 2004. "The World Social Forum and the Challenges of Global Democracy." *Global Networks* 4(4): 413–421.
———. 2008. *Social Movements for Global Democracy*. Baltimore: Johns Hopkins University Press.
Smith, Jackie, and Hank Johnston (eds.). 2002. *Globalization and Resistance: Transnational Dimensions of Social Movements*, edited by Jackie Smith and Hank Johnston. Lanham, MD: Rowman & Littlefield.
Snitow, A., D. Kaufman, and M. Fox. 2007. *Thirst: Fighting the Corporate Theft of Our Water*. San Francisco: Jossey-Bass.
Solomon, Steven. 2010. *Water: The Epic Struggle for Wealth, Power, and Civilization*. New York: HarperCollins.
Sorensen, Georg. 2006. "The Transformation of the State." In *The State: Theories and Issues*, edited by Colin Hay, Michael Lister, and David Marsh, pp. 190–208. New York: Palgrave Macmillan.
Spronk, S., and J. R. Webber. 2007. "Struggles Against Accumulation by Dispossession in Bolivia: The Political Economy of Natural Resource Contention." *Latin American Perspectives* 34(2): 31–47.
Strange, Susan. 1996. *The Retreat of the State: The Diffusion of Power in the World Economy*. New York: Cambridge University Press.
Subramaniam, Mangala. 2006. *The Power of Women's Organizing: Gender, Caste, and Class in India*. Lanham, MD: Lexington Books.
———. 2012. "Grassroots Groups and Poor Women's Empowerment in Rural India." *International Sociology* 27(1): 70–93.
———. 2014. "Neoliberalism and Water Rights: Case of India." *Current Sociology* 62(3): 393–411.

Subramaniam, Mangala, and Beth Williford. 2012. "Contesting Water Rights: Collective Ownership and Struggles Against Privatization." *Sociology Compass* 6(5): 413–424.
Subramaniam, Mangala, and Laura Zanotti. 2015. "Introductory Essay: Environmental Justice-Just Livelihoods." *Politics, Groups, Identities* 3(4): 649–654.
Subramaniam, Mangala, Manjusha Gupte, and Debarashmi Mitra. 2003. "Local to Global: Transnational Networks and Indian Women's Grassroots Organizing." *Mobilization* 8(3): 335–352.
Swyngedouw, E. 2004. *Social Power and the Urbanization of Water: Flows of Power*. Oxford: Oxford University Press.
Swyngedouw, Erik. 2005. "Dispossessing H_2O: The Contested Terrain of Water Privatization." *Capitalism Nature Socialism* 16(1): 81–98.
Tabb, William K. 2004. *Economic Governance in the Age of Globalization*. New York: Columbia University Press.
Tarrow, S. 2005. *The New Transnational Activism*. Cambridge: Cambridge University Press.
Tarun Bharat Sangh. 2006. *Arvari Sansad*. Jaipur: Jal Biradari.
Tripp, A. M. 2000. *Women and Politics in Uganda*. Madison: University of Wisconsin Press.
United Nations. 2005. "Women and Water." In *Women 2000 and Beyond*. New York: Division for the Advancement of Women, Department of Economic and Social Affairs of the United Nations Secretariat.
Wagadu Special Issue. 2006. *Issue on Water*. Volume 3.
Watkins, Susan Cotts, Ann Swidler, and Thomas Hannan. 2012. "Outsourcing Social Transformation: Development NGOs as Organizations." *Annual Review of Sociology* 38: 285–315.
Weiss, Linda. 2003. "Introduction: Bringing Domestic Institutions Back in." In *States in the Global Economy: Bringing Domestic Institutions Back in*, edited by Linda Weiss, pp. 1–36. New York: Cambridge University Press.
Whiteley, John M., Helen Ingram, and Richard Perry. 2008. *Water, Place, and Equity*. Cambridge, MA: MIT Press.
Widener, P. 2011. *Oil Injustice: Resisting and Conceding a Pipeline in Ecuador*. Lanham, MD: Rowman & Littlefield.
Williford, Beth, and Mangala Subramaniam. 2015. "Transnational Field and Frames: Organizations in Ecuador and the US." *Research in Social Movements, Conflicts, and Change* 38: 37–67.
Yashar, D. 2005. *Contesting Citizenship in Latin America: The Rise of Indigenous Movements and the Postliberal Challenge*. New York: Cambridge University Press.
Zwarteveen, Margreet Z. 1997. "Water: From Basic Need to Commodity: A Discussion on Gender and Water Rights in the Context of Irrigation." *World Development* 25(8): 1335–1349.

CHAPTER 3

Contesting Water Rights from "Below"

The shifts in the ways in which water rights have been viewed, utilized, and managed across the decades can be viewed as transitioning across three stages as discussed in Chapter 1: the First Water Age, the Second Water Age, and the Third Water Age (Gleick 2010). The First Water Age covered the period in which humans relied on the unpredictable national hydrological cycle to take what water was needed and to get rid of wastes. The transition to the Second Water Age began when humanity started to outgrow the limits of local water resources and to intentionally maneuver the hydrologic cycle, building dams, irrigation canals, and wastewater systems, and putting in place the first laws and social structures for managing water. This Age covered the nineteenth and twentieth centuries, when modern water systems were increasingly put in place which brought benefits to societies, but were eventually inadequate as billions of people still lacked safe water and sanitation. At the same time, aquatic ecosystems continued to be ravaged by overuse and were contaminated (Gleick 2010).

Conflicts over shared water resources are growing when the globe is transitioning to the Third Water Age as increasing efforts are made to develop a sustainable system to use and manage this resource. It is in the Third Age (mainly after the 1980s) that water rights are couched within neoliberal thinking, intense protests and campaigns have emerged against the privatization of water, advocacy for community control of water (as a common property resource) has grown, and powerful campaigns against the bottling of water and bottled water have taken shape. There has been a significant shift in scholarly work on water-related issues from the First and Second Water Ages to the Third Water Age. While much of the early attention was directed at mechanisms for harnessing water resources and providing access to clean water primarily within the development studies literature, later work has focused on the struggles to control water as a public good and in the privatization of water supplies.

Campaigns against the privatization of water worldwide, from India, Bolivia and Chile to Canada and the United States, are testimony to the global scope of processes that tend to shift citizens and local people's control over resources such as water to corporate hands. Yet there are minor subtle differences in the forms of privatization being challenged across a variety of contexts. For instance, as discussed below, the analysis of the case of the Coca-Cola Company's attempts at bottling water at very low prices and high profitability in the southern Indian state of Kerala, among many other places, has drawn attention to the agrarian crisis and protests, the depletion of groundwater levels, and the contamination of water is an example of the effects of privatization (Raman 2010; Aiyer 2007; Shiva n.d.). Protests by local peasants led to the local *panchayat* (local governance institution) to cancel the license to the Coca-Cola Company (Shiva n.d.). The United States has also seen protests against the privatization of water. Consider, for instance, citizens' resistance to the sale of bottled water by corporations as existing water fountains are closed and new ones are rarely installed (Gleick 2010). Scores of battles against privatization are taking place in the developed and developing world, and the cases—two in India and one in the United States—examined in this chapter can be viewed as embedded cases within the broader struggles for control over water resources.

Context and Cases

Local groups and global forums have been organizing to find mechanisms for the collective management of water resources as well as for targeting local governments, national governments, and international financial institutions which are pushing for the privatization of water resources. Below I discuss three cases of people's struggles to actively resist privatization. Two of these cases are in India—Tarun Bharat Sangh (TBS) and the Plachimada struggle—and the third is the coalition of citizens in Stockton, California.[1] In the following section, I begin with a broad context of India and the United States and then provide the details of each case and compare the cases.

Context: India and the United States

In pre-independence India, the ruling British government functioned under the premise that the state owned water and that it had the authority to manage water as it deemed fit, without feeling obliged to involve the people in the process. This approach was contrary to principles of public trust to which England was committed for its own governance. Clearly, the resources of a colony were treated on a different footing from those of the rulers. According to the doctrine of public trust, people own water without formal title and the state holds water in trust for the people with fiduciary responsibilities. On the other hand, the colonial mindset was that the state, representing the Crown, had the authority to decide what was in the best interests of its subjects. According to Sankaran (2009), many discussions on sharing of constitutional powers in India have devoted attention to the sharing of powers among governmental organs rather than the sharing of power between the government and the people at large. If this is indeed so, the matter should be addressed at a constitutional level. There is incontrovertible scientific evidence that water in general, and water in India in particular, has to be managed for the common benefit, with participation by an informed citizenry capable of balancing rights with responsibilities.

Since independence, India has seen contentious control over water resources. It has gained momentum in the post-1980s when the economic reforms initiated put India on the path of liberalization with privatization as a major tenet. Pro-liberalization economists and international agencies have touted the benefits of privatization. But the Indian government has faced a variety of protests in its attempts to privatize the manufacturing as well as the service sectors. The national government was unable to ignore or overlook these protests because as a coalition government, the major political party was reliant on the support of a variety of smaller regional-based parties to stay in power. The balance of power at the national government changed in 2014 when the *Bharatiya Janata Party* (BJP) was swept to power following a major victory in the national elections. The national government has furthered the liberalization agenda by encouraging foreign direct investments and has fueled hopes of increased job opportunities. But these initiatives have been countered by labor unions, which have forcefully protested the turning over of their fate to the free-market principles espoused by the West and international institutions that are controlled by countries in the Global North. There has also been growing resistance to the alleged exploitation of human and natural resources, and concerns are being increasingly expressed about environmental justice. Access to food and water is one critical area where economic reforms have made themselves known in a major way for ordinary people.

The Indian government has formulated national water policies to address the water crisis (cf. Briscoe and Malik 2006). The first of these was in 1987, a revised version was issued in 2002, and a third and most recent one was released in 2012. The 2012 National Water Policy recognizes that the growing population and the impact of climate change can put further strain on the availability of water and raise the possibility of deepening water conflicts among different user groups. Among the many aspects of water raised in the recent water policy document, three are relevant for the discussion in this book: the responsibility of state institutions—at the national and sub-national levels—to ensure access to potable water for all citizens; pricing water to ensure its efficient use and reward conservation; and the provision of powers for water users associations to retain a portion of water charges and maintain the distribution of water in their jurisdiction. Users are people who are treated as beneficiaries and are required to pay for what were often viewed locally as common resources. Beneficiaries are defined by the state as recipients who are responsible to pay for services received, either through planning, management, financing, implementation, or all of the above. This theme is evident in India's national Five-Year Plans (Subramaniam 2014). While the 2012 water policy document lays out the intentions of the government, there has been no effort to translate these policies into action. Moreover, water itself is a state subject in India, that is, decisions regarding water are made by state-level authorities.

Many scholars, scientists, and policy makers note that India's water situation has to be addressed with great urgency (cf. Briscoe and Malik 2006). But there are disagreements in terms of how this may be done. Some advocate for a campaign of active public awareness, arguing that the powers in government will act only in response to public pressure. Others envision a constitutional recognition of water rights and water science as a basis for formulating water laws, statutes, and regulations that will provide a basic structure to predicate judicious and equitable actions for all citizens (Narasimhan and Gaur 2009). The shortcoming of these approaches is that water is a complex natural phenomenon with human as well as institutional dimensions. The central question to resolve is whether, in a democracy, the state's ownership of water is synonymous with people's ownership of it, or whether the state and people are different. People's struggles for managing and controlling water resources is therefore concerned with the power of both the state and the private sector in gaining control over water resources (Gujja et al. 2008; Jha 2002; Kaur 2003). This overview of water politics in India may vary somewhat between the two Indian cases as they are based in two different states; Kerala and Rajasthan.

Kerala is a state in southern India with a population of 33,387,677 as per the 2011 Census (the thirteenth-largest in India). A little over 1% of the population are Scheduled Tribes or Adivasis (Bijoy 2006, 2008). Around the same time as the setting up of the Coca-Cola plant, the Kerala government, responding to a statewide Adivasi revolt, promised that all landless Adivasis and those owning less than an acre would receive up to five acres within one year.[2] This did not happen. However, the permission granted to Coca-Cola to set up the plant was viewed by the state as a means to develop this "backward" region. Yet Kerala has the highest Human Development Index (HDI) (0.712 in 2015), the highest literacy rate (93.91%), and the lowest infant mortality rate (see Table 3.1). Kerala hosts two major political alliances: the United Democratic Front (UDF), led by the Indian National Congress; and the Left Democratic Front (LDF), led by the Communist Party of India (Marxist) (CPI(M)). Since 2016, the LDF has been the ruling coalition in power. Drawing on a tradition of popular mobilization, a number of non-governmental organizations (NGOs) have been seriously challenging both the state government and business interests since the 1970s. The focus of many of the new movements was on the issues of land,

Table 3.1 Main indicators for Kerala and Rajasthan

State	Literacy rate (2011)*	Sex ratio at birth[a] (2014)	Infant mortality rate (2013)*	Maternal mortality rate (2011–2013)*
Kerala	93.9 (1)	948 (2)	12 (1)	61
Rajasthan	67.1 (18)	799 (19)	47 (15)	244
India	74	887	40	167

*As reported in Sinha (2016) Ranking among 19 states for which data are available is in parenthesis
[a]As reported in Vital Statistics of India Based on Civil Registration System 2014 (Office of Registrar General 2014)

water and the natural environment. The best-known environmental movement is the Silent Valley movement which successfully protected larger areas of ecologically sensitive forest from being affected by the construction of a nearby hydropower dam (Berglund n.d.). This grassroots resistance was organized without the support of any political party. Since the Silent Valley movement, a number of similar campaigns have arisen—the Plachimada struggle discussed below is one of them.

In contrast to Kerala, Rajasthan is behind most other states of India in terms of education and gender-related indicators. The literacy rate at 67% is lower than Kerala's (female literacy rate in Rajasthan is the lowest in the country) and infant and mortality rates are higher than the average for India. Poverty rates are high in Rajasthan compared to Kerala. The state of Rajasthan has not seen active protests or organized movements as those in Kerala. NGOs have been active in providing services such as education and healthcare, and indirectly facilitating the mobilization of local residents to access their basic needs. It is such NGO-led activity that has somewhat enabled the organizing of village residents in the Alwar district of Rajasthan.

In developing countries like India, there is also a pressure to privatize because of the conditions put on loans by global financial institutions such as the World Bank (Gies 2009). Overall, people's protests against privatization are connected to the fear of losing access to basic resources and of being left at the mercy of the free-market system, with businesses and corporations hardly being held accountable for their lack of scruples. The successful outcomes of these protests frequently led the state to adopt partial privatization policies, often compelled by the authority exerted by various state institutions across the sub-national levels. Thus, markets are not systems or entities isolated from the social dimensions of power. These developments in India differ greatly from those of the United States, which has pursued the Anglo-American model of capitalism and privatization.

In the nineteenth century, water ownership and management in the United States was largely in private hands. For example, in New York, the city's private water provider, the Manhattan Company, was actually a front to win public subsidies. Company founder Aaron Burr then diverted the money to establish a bank to compete with the Bank of New York, which was controlled by his arch-enemy, Alexander Hamilton, who Burr later killed in a duel (Snitow and Kaufman 2008). Although Burr's Manhattan Company was a disaster as a water company, it survived as a bank, becoming Chase Manhattan, buying out Hamilton's old Bank of New York, and finally becoming JP Morgan Chase. The company's logo, however, remains the same: a cross-section of a nineteenth-century wooden water pipe (Snitow and Kaufman 2008).

The story was similar in cities across the United States. As populations grew, private companies did not have the resources or expertise to meet the growing needs. Citizens demanded and eventually won modern public water systems, financed through bonds, operated by reliable engineers and experts, and accountable to local governments. The nation built a great system of community waterworks that provided clean, reasonably priced water and sewer systems that still rank among the best in the world. But in recent years, the federal disinvestment in water services has sparked a new era of privatization with contemporary players repeating promises made by nineteenth-century entrepreneurs. The radical shift to the private sector could become one of history's largest transfers of ownership, control, and wealth from public trust to the private till. But there is more at stake. The concept of democracy itself is being challenged by multinational corporations that see Americans not as citizens, but merely as customers.

Water is a huge market. Public utilities in the United States are used as a model in many parts of the world. Public operations ensure transparency and provide communities with the opportunity to work for positive action through public hearings and citizen action (Snitow and Kaufman with Fox 2007). About 85% of Americans receive their water from public utility departments, making water infrastructure, worth trillions of dollars, a prime target for privatization. To drive their agenda, water industry lobbyists have consistently opposed federal aid for public water agencies, hoping that federal cutbacks would drive market expansion. So far, the strategy has worked. In 1978, just before the Reagan-era starvation diet began, federal funding covered 78% of the cost for new water infrastructure; by 2007, it covered just 3%. In the 20 years ahead, local and state governments, ratepayers, and taxpayers will have to come up with well over half a trillion dollars for drinking water purification and sewage treatment (Snitow and Kaufman 2008).

In recent years following a worldwide trend toward privatization across sectors, private companies have started assuming the management of municipal water systems in countries around the world. Capitalist models argue that private firms are more efficient and effective than government-run entities, and this rationale can be compelling to local governments in developed countries (Gies 2009). Following this trend, the world's largest private water companies have quickly entered the American market: Suez and Veolia from France and Germany's RWE/Thames. Although few Americans have heard of them, even though their reach has been wide. Veolia purchased U.S. Filter in 1999 and Suez acquired United Water. Two years later, RWE's subsidiary Thames Water purchased American Water Works, the largest US-based private company, taking on $3 billion in debt in the process. The companies moved quickly to gain market share. Between 1997 and 2002, the number of municipal water contracts with the private industry tripled. In fact, Food & Water Watch, a US and Germany-based consumer advocacy organization, has even tracked the rate increases after privatization in some states such as California and found that privately owned utilities charged customers up to 50% more than those that were publicly owned (Gies 2009). Citizens began to resist privatization as they became aware of the adverse consequences of such privatization (Snitow and Kaufman with Fox 2007).

Resistance to privatization has grown across states in the United States In Lee, Maryland, the revolt was against potential water plant layoffs. In Felton, California, it was about rate increases and local control. In Atlanta, it was broken pipes and sewage lines. In other communities, it was the fear of corruption, cover-ups, and complicity between politicians and giant corporations. An epicenter of this nascent movement has been Stockton, California, whose population of 250,000 is at the heart of the state's agricultural San Joanquin Valley. As discussed in Chapter 1, a citizen's group in Stockton took on not only the mayor and the city council but also some of the world's largest private water corporations in a preview of the corporate water wars to come (Snitow and Kaufman 2008). In fact, the citizens groups discussed in this book (see the overview given below) and the cases from India are about resisting privatization and demanding control over water resources.

Let us now turn to discuss the three cases: TBS in Rajasthan, and the Plachimada struggle in India and the Concerned Citizens' Coalition of Stockton (CCCS) in California.

Cases: Local Struggles over Water

The mobilization of the local people in north-western Rajasthan, India was facilitated by an external organization, TBS, led by Rajendra Singh, known as the "waterman," who has since become a nationally and internationally known activist supporter of community rights to water and against the privatization of water resources. It was in 1985 that just five young men of the organization, along with its secretary, Rajendra Singh, came to the Kishori village in Thanagazi block of the Alwar district in Rajasthan, a north-western state of India.[3] At that time, there was little grass for grazing cattle and crop yields were very low. Just 3% of the cultivable area was irrigated. Around 90% of villagers were marginal farmers with their own land. Migration was acute. In such a grim situation, the TBS intervened as per the advice of the locals by restoring the traditional water harvesting structures called *johads* (earthen check dams) in the region to capture and conserve rainwater, which improved percolation and groundwater recharge. Government authorities opposed this organization as it directly interacted with the people and mobilized them for the work. But TBS continued to work with the mobilized community and organized them to consolidate the initiative. Its greatest symbols are five rivers of the region, which have started flowing perennially after decades of drought, a direct result of conserving water in *johads* (Mahapatra 1999) (Table 3.2).

By the end of the 1990s, TBS had expanded its work to 10 districts of Rajasthan (Menon et al. 2007). It is claimed that by 2005, TBS had worked with the rural communities of 700 villages and had built thousands of water-harvesting structures in various parts of Rajasthan (Menon et al. 2007). In terms of the number of villages covered and structures built, the scale of TBS' work could potentially be compared to that of state agencies. In April 2001, TBS with Rajendra Singh as

Table 3.2 Comparison of cases

Cases	Tarun Bharat Sangh	Plachimada struggle	Concerned Citizens Coalition, Stockton
State/Country	Alwar district, Rajasthan (India)	Plachimada, Pallakad District, Kerala (India)	Stockton, CA (United States)
Location/Place	Rural-based	Semi-rural	Urban-based
Key goals	Consolidate local water resources Control over water resources by village residents	Shut down the Coca-Cola water bottling plant, which had reduced groundwater levels and contaminated the water	Resist privatization of city's municipal water utility

the leader started the "Jal Biradari" program to raise awareness about India's National Water Policy among people. In 2012, as discussed in Chapter 1, Rajendra Singh and his TBS launched "The Flow Research Network" intended to be a research network linking global communities to enable the sharing of details of concerns of local citizens and their connection to water.

Unlike the TBS, the Plachimada struggle, in the southern Indian state of Kerala, comprises a loose network of local citizens without a formal organizational structure. The Perumatty *panchayat* (the local governance institution) granted a license to the Coca-Cola Company in January 2000 and the plant was commissioned in March.[4] The struggle began when residents of the village began to experience a depletion in their groundwater resources as well as the pollution of their water supplies. The residents soon realized that Coca-Cola had appropriated the village's "common" water resource for its water bottling plant. When it was initially set up, Coca-Cola had received a conditional license from the *panchayat* to draw a limited amount of water. But the company started extracting "millions of liters of clean water" (Shiva n.d.: 2) resulting in a drop in the groundwater level, going from 150 feet to 500 feet. Moreover, the company was pumping waste water into the dry borewells as a means of disposing of solid waste. The local *panchayat* served a show cause notice and canceled the license when the company failed to respond to their request for details.[5] Coca-Cola offered to pay the *panchayat* President, who declined to be co-opted.

In 2002, almost 300 people were arrested at a mass rally at Plachimada protesting the excessive use of groundwater by the Coca-Cola bottling plant. In 2003, the District Medical Officer declared the water unfit for drinking.[6] "Coca-Cola had created a water scarcity in a water abundant region" (Shiva n.d.: 2; see also Jayaraman 2002). Mostly Adivasi (tribal) women staged a *dharna* (a sit-in) and on Earth Day 2003, they organized a huge rally. The *panchayat* also filed a Public Interest Litigation against Coca-Cola in the state High Court (Chekutty n.d.). The Court ruled in favor of the women and ordered Coca-Cola to "stop pirating Plachimada's water" (Shiva n.d.: 4). The Plachimada protestors relied on a legal path to address their concerns (Krishnakumar 2004a, b, c). On June 1, 2005, a Division Bench of the Kerala High Court asked the Perumatty Gram Panchayat to issue a license to the

Coca-Cola Company at Plachimada. Having been legally advised to first comply with this order and then move the Supreme Court against it, the Gram Panchayat issued the permission for three months with 13 conditions on June 6. The Anti Coca-Cola Struggle Committee decided on June 3 to not let the Plant function; 300 agitators were arrested six days later (they were subsequently released). Meanwhile, the National Alliance of People's Movements (NAPM) continued its indefinite dharna before the Government Secretariat in Thiruvananthapuram (Kerala state's capital city), demanding the permanent closure of the plant. Subsequently, the state Chief Minister ordered the closure of the plant following intense pressure from the movement (Bijoy 2006; Pillai and Sreemadhavan 2008; Shree n.d.). Similar protests against privatization of publicly owned water resources occurred in Stockton through a formally structured citizen group (Snitow and Kaufman 2008).

In Stockton, CA, a strong mayor proposed awarding a $600 million, 20-year contract, to OMI-Thames, Inc. to manage and operate its municipal wastewater utility (Food & Water Watch 2007; Snitow and Kaufman with Fox 2007). For four years, the CCCS waged a grassroots campaign that culminated in a legal victory to defeat the privatization of their municipal water utility. The company was required to return control of the utility to Stockton, which became effective in March 2008 (Snitow and Kaufmann with Fox 2007).

Stockton's strong mayor played hardball in an effort to win approval of the largest privatization contract in the West, while a grassroots coalition went to the ballot box and the courts to stop him.[7] The CCCS was formed in 2001 to monitor and challenge what it called the Stockton city Mayor Podesto's push for privatization of the municipal water supplies. Podesto had a successful business career running grocery stores, which he sold out to a major chain store and retired comfortably. With a lot of time on his hands and with experience in dealing with the government, he decided to run for the position of mayor of the city. He brought his business ideology into this elected position, the primary motive being profit. His business model, built on privatization, formed the basis for managing the city's water system.

The Mayor's model immediately set up a conflict with employees like McDonald who worked at Stockton's sewage treatment plant, where he did almost every job for 26 years. McDonald believed that their system was not there to make profits, but to provide a basic human right. However, Podesto soon realized the perils of using the terms private and privatization and therefore adopted the more appealing phrase "public–private" partnership. In doing so, Podesto argued that "a private company running the water department would not be privatization because the city would still own the water and the facilities" (Snitow and Kaufman with Fox 2007: 28). Members of the CCCS disagreed because such a partnership is privatization when a municipal function is turned over to the private sector. The major mission of the CCCS became the fight against the privatization of water.

In 2003, the City Council of Stockton ignored overwhelming public opposition to approve a $600 million, 20-year water privatization agreement. The deal gave a multinational consortium full control over the city's water, sewage, and stormwater systems. But two weeks later, the Council reversed its position and voted unanimously to resume control of its water utilities.

The Mayor pursued his agenda. Analysis by consultants brought in by him claimed that the city would save $175 million over 20 years by contracting out water services. The coalition drew on the expertise of an Oakland-based think tank, the Pacific Institute for Studies in Development, Environment, and Security, to counter the suggestion of the consultants. In its independent assessment, the think tank noted that the savings had been over-estimated and that it would cost less to continue the public operation rather than bring in a private company. The Mayor was unwilling to consider these findings, which were put forward by the CCCS. He also saw a problem with the Director of the Municipal Utilities Department, Morris Allen, who did not support the Mayor's goal of privatization. Allen, along with other supervisors, had already streamlined operations, which contributed to creating substantial savings for the city. Although some City Council members were in favor of the continuance of the streamlining process, the Mayor declined to give up on privatization.

Focusing on these three cases, I discuss below four main aspects related to collectively or publicly managing water resources as a right of local communities: local mobilization to organize collectively; relying on indigenous knowledge for the collective ownership of water resources; strategies used to challenge state institutions; and the ways in which the intersections of gender and/or caste and/or class are constitutive of local power relations in accessing and controlling water resources.

The Ambivalent State and Struggles Against Privatization

Struggles against the privatization of water involve state institutions at the national and sub-national levels. Without any formalized rules, people have established and enforced rights to water for hundreds of years. These rights vary according to time, place and individual situations, but they all demonstrate indigenous people's ability to manage their own water systems. Local citizens opposed to privatization are concerned that, once privatized, water will no longer be provided based on need, but rather on the ability to pay. Many people across the world are not able to pay for water, so the question arises of who privatization aims to help. Economists further use ambiguous terminology such as "appropriate" and "efficient," and refer to the cost of water, value of water, and water efficiency in their discourse (Meinzen-Dick et al. 1997; Bauer 1997; Brewster et al. 2006), raising further questions of who determines what is an "appropriate" amount of water and for whom. Not all values are as quantifiable as economists assume, especially when they are embedded in culture and identity. These concerns have mobilized local citizens in villages and cities to organize and seek to make water a public resource.

Mobilizing Locally

In all three cases discussed in this chapter, the concern for ensuring water availability motivated local residents and citizens to mobilize and manage water resources. Participation in public decision-making and collective activities has very obvious costs and benefits in terms of time and effort. Such costs and benefits affect men and women differently. For the poorest people, the opportunity costs of such participation may be prohibitive (United Nations 2005). These notions of costs and benefits are also tied in with considerations about the control and management of water resources from an economic perspective, which focuses mainly on questions of equity and efficiency. Equity and efficiency overlook considerations of motivations beyond individual rationality as involving relationships and emotions; a sense of collective identity possibly linked to a shared history of struggles to maintain livelihoods. The diversity in local users' interests and respect for the value of water reflects the broader equity principle of solidarity, which includes recognizing that each person or group has a right to participate in open and inclusive decision-making processes so that their values and visions can be considered fairly (Priscoli et al. 2004). Social differences based on gender, class, and caste as well as the strategies adopted had some influence in all three case studies of struggles considered here.

A significant event which facilitated the initial mobilization of the local community in the Alwar district is the rejuvenation of the Arvari River and the conflict that followed over whether this river water and the fish therein belonged to the government or the community. In 1996, the government issued rights to the fish to a private contractor for a small amount of 18,700 Rupees (1 US$ = 45 Rupees approximately), although the market value was over 100,000 Rupees (Tarun Bharat Sangh 2006). The community residents protested against this and decided to keep vigil over the river so that the contractor would not have access to the river. A conflict followed between the community, the local government department of fisheries, and the contractor. The pressure from the community led the government to annul the contract for the fish. The coming together of the village residents to protect the Arvari River was an assertion of the rights of the community.

By setting up *Lok Samitis* (community organizations), TBS mobilized local village residents to participate in saving and managing water resources. For instance, the *Samitis* directly implemented the Pavdi watershed initiative of the Government of Rajasthan and also started to take up the issues of mining in protected areas, child labor, education and health. In addition, the village assembly, which consists of members of every family in the village, played an important role in ensuring the wide participation of villagers in decision-making and the implementation of activities. In 1992, the role of assemblies was reaffirmed by the government. Representatives from the Assembly are designated members of the *Arvari Sansad* (Arvari is the name of the river revived by this community and *sansad* means "parliament"). The parliament meets twice a year; generally in June and December, and serves as a forum for discussion of the three issues central to this movement: trees, forests, and water. For instance, the December 2007 session included a discussion of whether farmers who have the resources should be boring for water, as groundwater levels will fall. Through the water parliament system, TBS has stressed self-governance. By creating rules and limits on water use, collective water management may ensure sustainability and equity.

A combination of formal structure with informal localized organization can also facilitate participation.

Community understandings of participation differ from those of the state, particularly as local people look to determine the rules and norms for using and saving water. In fact, invoking participation entails mobilizing and organizing; an involvement in the processes of creating water-related structures, their maintenance, and mechanisms for regulating use. This has been well articulated by environmental activist Anil Agarwal, the late Director of the Centre for Science and Environment (CSE) in Delhi:

> Building water harvesting structures is a very easy task—any contractor endowed with a bit of money can do so. But building an effective structure which starts off a process of self-management in village communities is a much more difficult task. This is possible only if each structure is the result of a cooperative social process—the ability of a community to work in cooperation. (Agarwal 2001: 13)

Noting that water is a "strange" natural resource that can unite or divide a community, Agarwal emphasizes the need for the strong social process that must precede each structure to build social capital. Interestingly: "[T]his is an area where the track record of government agencies is literally non-existent and inflexible government rules militate against the very principle of social mobilization" (Agarwal 2001: 13). Such social mobilization was among the first activities of TBS, which, as noted above, was opposed by the government.

The locally mobilized community in the Alwar district believed in protecting the rejuvenated Arvari river as they prevented the private contractor from gaining access to the fish and then contributed labor for the construction of small check dams to store water. But neither TBS nor the community assemblies collaborate with either government departments or the local *panchayat*. While TBS has successfully mobilized the community to regulate the using and saving of water, whilst simultaneously paying attention to the related issues of soil conservation and forest management, the 2007 parliament session was gendered, in that women had little to say or contribute to the deliberations. Women were present as passive actors, despite the fact that collecting water and using water is typically women's work. So, realistically, it must be acknowledged that local communities are constituted around social politics of gender, caste, and class which influence decision-making processes.

The CCCS functioned as an informal network. The network included a variety of local organizations/groups: environmental groups, labor unions, churches, NAACP as well as police and firefighter organizations. The CCCS was a congenial group of like-minded, mostly middle-class people from Stockton. It was chaired by Sylvia Kothe, an activist associated with the League of Women Voters. In addition to the support from Public Citizen, the national organization challenging water privatization and resisting corporate power, many employees from the city's utility department were active within the CCCS. But the CCCS was not about radical activism, which was evident from the lack of a response from the group to a suggestion by a young Filipino activist to pursue the policy of civil disobedience (Snitow and Kaufman with Fox 2007). The CCCS feared that the Mayor would push the privatization plan through without a thorough public hearing and a city-wide vote, and their research showed that the utility could be fixed without privatization.

The resolution to privatize was initially approved by the City Council by a vote of four to three. A push to let the public decide by voting ended with almost 60% of votes approving the initiative to require a public vote on a major change in the management of the municipal water supplies. The CCCS members devoted a considerable amount of time in meetings, making phone calls, writing letters, talking to friends and neighbors, and raising funds (Snitow and Kaufman with Fox 2007). The CCCS decided to gather signatures again for a city-wide referendum in order to overturn the City Council decision. But they decided to rely on volunteer signature gatherers, in part to save money and in part to ward off the Mayor's criticism of using paid volunteers. But the CCCS did not have enough committed volunteers to go out and collect signatures, and they did not foresee the difficulty they would face in getting signatures. The decision to use volunteers was debated within the CCCS as members expressed their concerns over the consequences of the decisions. For instance, Boisa, a State Assembly aide who worked to hire volunteers, spoke of the need to focus on the results (and win). Bagaloyas, the youngest activist, sought more radical action as well as the inclusion of minority groups, such as the growing Latino population.

Unlike the CCCS and TBS, the Plachimada struggle has involved the conservation of existing resources and protests against the excessive utilization of groundwater by Coca-Cola. The Kerala State Pollution Control

Board (KSPCB) gave Coca-Cola a permit to produce 561,000 liters of soft drink per day, requiring 3.8 liters of water for a liter of soft drink. In the Plachimada case, people, particularly Adivasis, came together, as they saw and experienced the depletion of their water resources, without any formal organization or leader. In the autumn of 2001, the first collective protests were organized outside the factory, demanding its closure and full compensation for the villagers (Berglund n.d.). In April 2002, the campaign against Coca-Cola was formally launched with the formation of the *Coca-Cola Virudha Samara Samithi* or CCVS (Anti-Coca-Cola People's Struggle Committee) and a blockade by over 1300 people, mostly Adivasis, demanding the shutdown of the plant as well as action against Coca-Cola for the destruction of the environment (Bijoy 2006). They were taken into custody and removed by the police. Many peasants, academics, environmental activists, youth, and Gandhians from across the state supported the struggle.

In fact, Coca-Cola had received aid totaling 2 million Rupees from the Kerala state government's industrial policy (Shiva 2005; see also Raman 2005, 2007). But unlike the state government, the state High Court, local opposition political party leaders, and the local *panchayat* responded in favor of the people's struggle. This raises questions about the state policy of participation in managing and maintaining water resources, as the people of Plachimada had not been consulted in relation to the establishment of the private factory; however, the local *panchayat* as the local government eventually pursued a policy that differed from the Kerala state government and the national government's agenda of attracting and supporting private investment of all forms.

Private investment is encouraged by neoliberal policies. Defining water as an economic good is tied to privatization in two ways: the first is contracting water harvesting structures (such as setting up tube wells and constructing check dams) to private enterprises and the second is providing private companies with access to water, often through special schemes under state industrial policies intended to attract investment. Both forms of privatization fail to actively engage village residents.

In sharp contrast to state policies driven by a privatization agenda, TBS actively engaged village residents in the construction of water harvesting and storage structures; part of this was treated as volunteer labor (locally referred to as *shramdan*) and part was paid labor. This mechanism facilitates the coming together of people to discuss the successes of water harvesting, when and how the structures will be built, who will build them, and how costs would be shared. After the structure is built, details of how the water will be shared, particularly early on when water is scarce, and how it will be regulated are significant to the processes of solidarity building of what may be only an informal community group. Anil Agarwal (2001) argues that at least the first two years of any water harvesting program will have to be spent on social mobilization. This also puts the onus of maintenance and regulation on the community. As noted by Agarwal: "Contractors must be totally kept out and all wage benefits should go to the landless" (2001: 14). In other words, this can serve as a temporary form of wage labor for the landless, particularly in rural areas.

Another form of privatization promoted by the state is one through which private companies have freely accessed groundwater for bottling or for industrial use, such as the case involving the Coca-Cola Company and the granite manufacturing industry. The industrial policy of the Indian state has increasingly emphasized privatization within the ambit of the neoliberal regime. In order to promote industrial growth, the national and state-level governments have created special schemes to promote private investment. This includes the provision of land, electric power, or water at subsidized rates and it is under one such scheme that the Kerala state government provided aid to Coca-Cola to set up a bottling plant. As discussed in Chapter 2, the demand for bottled water as the means to access safe and clean water has pervaded the developed and developing world. Multinational corporations have tapped water resources across the world and often indiscriminately, which have been met with protests. Yet as water sources are polluted, people are compelled to consider bottled water as safe, further allowing for the commodification of water. These issues are also at the heart of the conflict in Plachimada.

State policies enabling multinational investment were resisted by the local people of Plachimada as they collectively sought to control the use of water while trying to redefine the "water as a commodity" argument. In fact, the Plachimada Declaration notes that: "Water is not a commodity. We should resist all criminal attempts to marketise, privatise and corporatise water. Only through these means can we ensure the fundamental and inalienable right to water for the people all over the world" (Shiva n.d.: 15). These rights were recognized by the local *panchayat* and in the initial ruling by the courts, but it took continuous pressure by the movement to compel the state government to revoke Coca-Cola's bottling license. "The right to conserve, use and manage water is fully vested with the local community. This is the very basis of water democracy" (Shiva n.d.: 15). Thus, both the Arvari Sansad and the Plachimada movement emphasize the need to define water as a common property resource and the right to water as a human right, rejecting the economic commodification as well as any form of public–private partnership in managing water resources.

The public–private partnership paradigm was also rejected by the CCCS, which argued that contracting municipal water supplies to a private company was tantamount to privatization. The phrase "public–private partnership" is the central rhetoric of neoliberalism advocated by the World Bank and the IMF. The motive of profit remains intact in public–private partnership arrangements and the public have no mechanism for participating in decisions related to the management of the municipal water supplies.

The motivation to conserve and regulate the use of water and arrest its depletion facilitates the mobilization of residents and citizens. The commodification of water, a process by which water is perceived less as a social good and more as a marketable commodity, to be bought and sold, was resisted by people in all three cases. Local groups have managed water resources by utilizing indigenous knowledge for harvesting, managing, and saving water. The "water knowledge" held by indigenous groups is embedded in many cultural norms and traditional community practices. Few scholars have made attempts to recognize and understand indigenous knowledge of water and water systems, and this knowledge is largely ignored in the macro-level discourse on water today. This is because as water has become increasingly scarce and privatized as a commodity, the need for an external agency or institutional monitor has been increasingly felt. The indigenous knowledge in managing/saving water and understanding water rights has been unrecognized in modern-day water debates. In comparison, the mostly educated participants in the CCCS were keenly aware of the pitfalls of privatizing the city's water supplies and the long-term implications for prices and quality. The level of knowledge and protecting public water utilities as the goal of CCCS differed from the cases of TBS and the Plachimada struggle, which are based in rural and semi-rural areas respectively.

Local Knowledge for Collective Ownership

Many indigenous people of the world view water as embedded in their cultures, as a sacred element, and people as holders of "water knowledge." Indigenous people are often quoted as saying "water is life."

Thus, in contrast to the view of the state (and private corporations) that water is a commodity, local citizens argue that it is a common property resource which can be managed using local innovation and indigenous knowledge. "Water knowledge" is embedded in traditional local practices such as collecting and storing rainwater for domestic and agricultural use. For instance, some like Agarwal (2001) in India have argued that rainwater harvesting at the village level supports local people-led conservation and regulation.

The attention to traditional forms of water management is tied in with the discussions in other water-related forums. Responding to the drought in the Gujarat and Rajasthan states in 2001, Anil Agarwal noted that the "solution lies in harvesting rainwater through capturing, storing and recharging it and later using it during prolonged parched periods" (2001: 8). Agarwal argues that there is scientific evidence to show that rainwater harvesting at the village level will yield more water than large or medium-sized dams, making the latter cost-effective and raising the possibility of community-led conservation and regulation. This relentless campaign by the CSE Director backed by scientific evidence led political leaders to respond to the national debates on the drought by making statements "regarding the importance of a community-based rainwater harvesting strategy to drought-proof the country" (Agarwal 2001: 11).

Through TBS, the Arvari community has utilized traditional water-harvesting structures, *johads*, and also adopts cultural and religious practices in conducting environmental campaigns. Between 1985 and 2001, about 4500 water-harvesting structures were reported to have been constructed in 850 villages in Alwar (Menon et al. 2007). While the local protestors in Plachimada have not constructed *johads*, they objected to and resisted the more elaborate and expensive system of boring of wells which Coca-Cola purportedly indiscriminately engaged in. Boring of wells directly impacts groundwater levels; falling levels will make water more and more difficult to access. Communities have knowledge of local topographies and the benefits of smaller structures based on their experience, and this was long before the government acknowledged and recognized the usefulness of smaller water harvesting structures after the drought in 2000–2001.

In the case of the CCCS, largely educated and mostly middle-class men and women (Snitow and Kaufman 2008) who were well aware of privatization of water supplies protested against the city council's actions. As noted above, it is important to note that the CCCS group members were familiar with municipal water supplies and what that meant for the city's population, and are not specific marginalized groups as in the case of TBS or Plachimada. While indigenous knowledge for saving and using water was not a major emphasis of the CCCS in Stockton, the strategies they adopted to reverse an initial City Council vote involved time, effort, and money.

Strategies Adopted: Local and Global Alliances

In each of the cases discussed in this chapter, there is some overlap in the strategies adopted, albeit with some differences (see Table 3.3). For instance, both the CCCS and the Plachimada struggle strategically considered a legal path by appealing to the courts and thereby also targeted a state institution independent of the legislative branch to meet their goals. The court system was available to the people in the United States and India, both of which are democracies. The people in Plachimada also targeted the *panchayat* to withdraw the license issued to Coca-Cola (Bijoy 2006; Koonan 2010). In the case of TBS, strategies involved sharing knowledge and information about water policy with local residents and enabling discussions about local water resources, particularly the different uses and conflicts (if any), through the "pani sansad" (water parliament). The Plachimada protestors made efforts to build global alliances, as Coca-Cola is a multinational company and has faced similar protests in other places (see Chapter 4). In fact, in Plachimada, the protestors organized local dharnas (nonviolent sit-in protest that may often involve fasting), blockades, and marches or rallies. But the CCCS relied on the required signature campaign and utilized volunteers to push the Mayor and the City Council to withdraw the plan to privatize the city's water supplies. I discuss all these strategies below.

> What may be the most notorious water privatization deal in the United States came to a close July 17, 2007 when the Stockton city council decided against appealing a judge's ruling that its contract with OMI-Thames Water had violated California environmental law. For four years, the Concerned Citizens Coalition of Stockton waged a grassroots campaign that culminated in a legal victory to defeat the privatization of their municipal water utility. The company must now return control of the utility to Stockton, effective March 2008. (Food & Water Watch 2007)

Table 3.3 Comparison: struggles, state, strategies

Cases	Form of struggle	Participation	Strategies
Tarun Bharat Sangh (TBS)	NGO: working with village residents	Predominantly men; a few women attend, but do not participate in discussions	Developed collective strength among village residents to control local water resources Relied on local knowledge to build check dams for collecting water Organizes "pani sansad" (water parliament) to discuss water-related issues and share information about water policy
Plachimada struggle	Informal group: grass-roots-based, launched as the Anti-Coca-Cola People's Struggle Committee	Adivasis (tribal), men and women	Targeted local *panchayat* and drew global attention to pollution of groundwater by Coca-Cola, a multinational company Used legal means Local dharna, blockades, marches, and public rallies demanding the shutdown of Coca-Cola's plant Built global alliances
Concerned Citizens' Coalition of Stockton (CCCS)	Informal group	Primarily middle-class educated men and women Efforts to include groups such as the Latino population	Relied on volunteers to spread the word and work Used a signature campaign to challenge the Mayor and the City Council Used legal means

In the fall of 2002, the multinational consortium of OMI and Thames Water was chosen to negotiate a contract with the city, which was a 20-year deal worth $600 million. This was an attractive contract for this consortium, because this well-run utility service was at the center of water politics. Allen and other employees reviewed the contract, including the design plans, and raised concerns that led to a request to take the time to consider alternatives. But Podesto would have none of this and went on to strip Allen of his position as Director and sent him to Boston to meet with consultants. This led McDonald to announce that he would not work for a private company if that contract went through. Moreover, Podesto decided to hold a vote in the City Council before the set date, taking the CCCS by surprise. The date for the vote was also the date of the annual State of the City address. In his address, Podesto called for entering the twenty-first century in terms of the provision of services and to think of citizens as customers.

City Hall was in pandemonium before the vote. Sewage plant workers and supporters rallied on the steps carrying banners such as "no to privatization" with chants of "let us vote" and "this is a democracy." The Mayor had set up the list of speakers which included the OMI President, Don Evans. The public hearing closed with Evan's assurances that OMI would safeguard the water of Stockton as well as making a personal commitment to do so. Members of the Council voiced their opinions; the majority were in favor of the privatization contract. The City Council voted by a majority of four to three to ratify the contract. Two weeks later, the public vote on the citizen's initiative renewed the hopes of the CCCS because 60% of voters approved the requiring of public votes on major changes to their water system.

The pressure from CCCS members, particularly at City Council meetings, made the Mayor limit his comments and also refuse to disclose the details of the privatization proposals until the winning bid was decided. Frustrated, the CCCS decided to go to the voters in an effort to get the initiative on the ballot. They had 90 days to gather the required signatures. Meanwhile, the Mayor was annoyed with the CCCS, but gave in by establishing an ad hoc committee on water privatization with members from the community who were mainly his allies. With few or no resources, the CCCS was able to gather more than 18,000 signatures within 90 days. The first vote was scheduled for March 2003, which was more than six months away.

The CCCS responded by organizing a referendum campaign to cancel the contract. After weeks of walking their neighborhoods, they fell just shy of collecting the required number of signatures to nullify the contract. They lacked a sufficient number of committed people or volunteers willing to go out and collect signatures, which in an internal discussion was somewhat divisive. Their efforts were complicated by a ruthless and expensive campaign launched by OMI/Thames with Mayor Podesto aimed at defeating the citizens' campaign. Left with no other alternative, the CCCS sued the City of Stockton for not conducting an environmental review of the contract as required by Californian law. In a case that garnered international attention because it was featured in the documentary film *Thirst*, the CCCS emerged victorious before and after an appeal by the city. Nevertheless, while the case was argued in court, the citizens of Stockton experienced three years of privatization.

Every month that OMI/Thames was in control of the water department, facts were being constructed on the ground. Water rates that had been stable for five years under public control rose every year following privatization, just as the CCCS had predicted. Mayor Podesto had claimed that rates would rise only 7% over the 20-year life of the contract, but a detailed CCCS analysis reported increases of 8.5% in just the first three years. Forced to retire because of term limits, Mayor Podesto ran for the state Senate, expecting to upset the Democratic incumbent. During the campaign, his opponent came out against the water privatization in Stockton. Podesto was portrayed as being on the side of corporate owners and developers, and lost the campaign. In a sense, the CCCS strategically drew in the public to vote at key moments in the process in order to reverse Mayor Podesto's and the City Council's decision. Like the CCCS, TBS and the Adivasis in Plachimada also approached the court at some point.

In the Alwar district, Bisleri and Coca-Cola factories have established themselves as major consumers of large amounts of water. Local communities protested and a Public Interest Litigation hearing is before the courts. In fact, Rajendra Singh along with other activists and protestors were arrested as they marched toward a bottling plant of soft drinks giant Coca-Cola in the neighboring district of Jaipur in December 2005 (News Hour 2008). The Plachimada protestors also approached the court to get a stay on the operations of the Coca-Cola and to shut it down. They used multiple strategies to draw attention to their cause.

The residents of Plachimada chose to build broad alliances to push for the closure of Coca-Cola's plant. A World Water Conference organized in January 2004 also brought international activists such as Jose Bové to Plachimada to support the protest. The vibrant state literacy movement, the well-known local newspaper *Mathrabhumi*, as well as political leaders specifically from the opposition (party not in power in the state) supported the protesters of Plachimada. The broad alliance put out the "Plachimada Declaration," which defined water as a fundamental right of all, as a common property resource which the community has rights over, and declared that water is not a commodity (Shiva n.d.).

Local groups in countries such as India have released written declarations that water is a common property resource. For instance, the ongoing Plachimada struggle involved broad alliances and multiple strategies to demand the closure of the Coca-Cola plant in the southern Indian state of Kerala. The residents organized dharnas, blockades, sit-ins, marches, public rallies, and meetings. The non-adivasi farmers living in the village who had not initially supported the movement altered their stance as the depletion of water levels was evident in the borewells owned by them. Yet although farmers with land holdings in the area of the factory who were affected by the water shortage supported the movement, they did not participate in protest activities (Berglund n.d.). The concern of these farmers seemed to be the lack of water rather than the pollution and environmental effects. The Plachimada Struggle Solidarity Committee took on the task of generating support at the national and global levels. By October 2008, 92 different organizations had registered as members of the Solidarity Committee (Berglund n.d.).

The Plachimada movement also formed ties with NGOs active in the areas of human rights, environmental issues, and anti-globalization movements at the global level. One such NGO was Greenpeace, which also financed one of the analyses of the contaminated wells of water. Another example was the World Water Summit held in 2004 to coincide with the World Social Forum in Mumbai and which included high-profile activists such as Vandana Shiva, Maude Barlow, and Jose Bové to Plachimada to support this struggle. They put out the "Plachimada Declaration" which defined water as a fundamental right of all, a common property resource that the community has rights over, and declared that water is not a commodity (Shiva n.d.).

In its statement against the proposal, Coca-Cola stated that it "recognizes that water is a precious natural resource under growing stress around the world." It set out the action that the company has taken to address the risks associated with water extraction and dealt with the complaints in Kerala.

As will be discussed in Chapter 4, concerns about Coca-Cola's factories were raised at the company's annual general meeting on April 19, 2006, in Delaware. The Board refuted the concerns and noted that in relation to groundwater issues in southern India specifically, the Kerala High Court ruling released in April 2005 (the result of a year-long independent study) stated that the facility was not the cause of water shortages in that community. The study showed that a cycle of three years of short monsoon seasons in the Kerala area was the main contributor to the local water shortages. The company referred to rainwater harvesting efforts in several communities and plant operations in India in a move to return a significant portion of the water removed from aquifers for production purposes. Additionally, it claimed to have initiated partnerships to set up local rainwater harvesting projects in communities around the country and to mobilize local residents behind these water conservation efforts. These projects combine modern technology with the reinstatement of traditional methods of water management that had fallen into disrepair in some local communities.

In Stockton, California, a strong mayor proposed to award a $600 million, 20-year contract, to OMI-Thames, Inc. to manage and operate its municipal wastewater utility (Food & Water Watch 2007; Snitow and Kaufmann with Fox 2007). For four years, the CCCS waged a grassroots campaign that culminated in a legal victory to defeat the privatization of their municipal water utility. The company was required to return control of the utility to Stockton, which became effective in March 2008 (Snitow and Kaufmann with Fox 2007).

The commodification of water by bottling it and the appropriation of both the process of bottling and selling bottled water are also being increasingly questioned by advocacy groups and activists. Instances of bottled water contamination are rarely reported to the public. These issues are not unique to the United States; similar controversies have arisen in India as well (News Feature 2007). In fact, the Coca-Cola plant in Plachimada produced both aerated and non-aerated drinks which required the use of water. The CSE, a lobby group, has actively engaged in the scientific testing of bottled drinks and water as water management is one of the five issues that it focuses on. For instance, in 2006, the CSE released a technical analysis of 12 popular soft drinks made by Coca-Cola and Pepsi Co that these companies made and sold in India, claiming that they contained toxic pesticides, including lindane, DDT, malathion and chlorpyrifos, at up to 36 times the European standards for bottled water (News Feature 2007).

Four states in India then banned these drinks (both the production and sale). But it has not all been straightforward. Coca-Cola and PepsiCo stood by their products, saying that their soft drinks made in India are of the same quality as those made anywhere else. The companies maintain that setting standards is complicated because there are no reliable tests to assess pesticide levels in carbonated beverages. Although the CSE data has not been peer-reviewed, scientists assert that the CSE followed methods accepted by agencies such as the US Environmental Protection Agency. These discussions have also led to PepsiCo's President, Indra Nooyi, announcing that the two multinationals would work to find a reliable set of measures to consistently measure the content of the finished products. And the CSE notes that they want standards and if they are wrong they expect the companies or the government to provide evidence. As these developments unfold, protests against Coca-Cola have grown globally (see Chapter 4). For instance, in the Alwar district, Bisleri and Coca-Cola factories have established themselves as major consumers of large amounts of water. Local communities protested and a Public Interest Litigation hearing is in the courts. In fact, Rajendra Singh, other activists and protestors were arrested as they marched towards a bottling plant of soft drinks giant, Coca-Cola, in the neighboring district of Jaipur in December 2005.

The Ambivalent State and Development: The Paradox of Jobs and Struggles

State institutions have been advocates of the neoliberal agenda by adopting privatization policies that have adversely affected environmental resources, including water. Local neoliberalism includes the contentious terrain of managing and using water, which involves the residents, NGOs, and state institutions. However, state institutions do not always work in concert with each other and may also alter their decisions over time and thereby appear ambivalent. This is also influenced by the lack of homogeneity among citizens (by gender, caste, and class) demanding their rights to water resources. The issue of how state institutions have been ambivalent in the protests and campaigns for water rights is what I discuss next.

The initial 2001 agreement between the government and the CCVS marked a watershed in the contemporary history of the Adivasi struggles (Bijoy 2008) because it was intended to provide jobs for the local poor. Touted as a development initiative by the government, the firm employed about 130 permanent workers and 250 contract laborers. This was particularly important because the average daily wage was low in this region—about 80 Rupees for men and 40 Rupees for women— and dwindling to about 120 days of work per year.[8] However, as noted above, within a short period of the start of the Coca-Cola plant's operations, the residents of Plachimada experienced problems that they had never encountered before—the receding of the water table and the drastic change in the quality of water. Moreover, water shortages had an adverse impact on agricultural operations and the deteriorating quality of water made it unfit for consumption.

The "socially responsible" Coca-Cola plant initially sold and later gave away for free the slurry and sludge waste as fertilizer to local farmers, who were unaware of the contents of the waste. As the residents began to recognize the problems, Coca-Cola surreptitiously dumped the sludge on the wayside and on lands at night. The residents questioned the form of "development" the Coca-Cola plant was engaging in the promotion of such privatization efforts by the state has implications for environmental justice. But consideration of the state was not entirely straightforward in this case as the power of various institutions at the sub-national level was exercised at different points of the increasingly popular struggle. Moreover, as I discuss next, sub-national state institutions changed their position on the Coca-Cola plant's operations as the struggle intensified.

During the initial two months, the struggle faced hostility and threats from local political parties, which included the Congress, the party in power in Kerala at that time, Janata Dal, the Communist Party of India (Marxist), and the BJP. As scholars have noted, there have been suspicions about the government being lenient on Coca-Cola when investigating accusations of pollution and water depletion (Pillai and Sreemahadevan 2008; Bijoy 2006; Berglund n.d.). For instance, early on in the campaign, when the CCVS was demanding the shutdown of the plant, Coca-Cola filed a case in the Kerala State

High Court in April 2002 against the struggle committee, demanding that the picket be dismantled and police protection be provided to the plant. The Court conceded the right of the people to protest peacefully and ordered the police to provide protection to both the plant and to protestors. However, the government provided heavy police protection for the plant alone (Bijoy 2006). Besides the state government, four key state institutions were involved in the dynamics in the Plachimada–Coca-Cola struggle: the Perumatty *panchayat*, the Local Self-government Department, the State High Court, and the KSPCB.

A key state institution was the local Perumatty *panchayat* controlled by the Janata Dal, which was initially hostile to the struggle of the people of Plachimada (Bijoy 2006). But as the struggle intensified, the *panchayat* was influenced by the Committee in 2003 to recall the production license issued to the plant (Berglund n.d.). The *panchayat* cancelled the license and ordered Coca-Cola to stop production from May 2003. A critical point to note here is that the *panchayat*, a people's institution, had not considered the negative effects before issuing the license. The Local Self-government, approached by Coca-Cola at the behest of the High Court, stayed the decision of the *panchayat* in June, stating that it had exceeded its powers conferred under the Kerala Panchayat Raj Act. But the Local Self-government ordered the *panchayat* to constitute an expert team for an investigation into the fate of the license given to Coca-Cola. However, the *panchayat* returned to the High Court, asserting its right to refuse the license to Coca-Cola, a point that was upheld by the Court. Coca-Cola challenged the Court's decision, but without success. The High Court orders became more stringent as it ordered Coca-Cola to stop extracting groundwater beyond what was required for irrigating 34 acres of land. The Court simultaneously required the government to investigate the allegations relating to the contamination of water and land. In its ruling, the Court noted that groundwater was public property held in trust by the government and that it had no right to allow a private entity (Coca-Cola) to over-exploit a natural resource—in this case, water.

The ongoing dynamic between the *panchayat*, the High Court, and Coca-Cola continued until early 2004, when Coca-Cola was granted one month to close down its wells and find an alternative source of water. Moreover, while the *panchayat* declined to renew Coca-Cola's license in 2003, the Kerala state government intervened to overturn the *panchayat*'s decision. But the Court stayed the government's decision and, on the directive of the Court, the plant stopped operations in March 2004. But Coca-Cola did not give up; it appealed to the KSPCB in September 2004. The Board noted that the Coca-Cola sludge contained cadmium above the tolerance limits. Moreover, in January 2005, protestors blockaded the factory, declaring that any reopening would be prevented. However, the battle between the *panchayat* and the High Court continued until the end of 2005 as the Court required the government to provide an investigative report. The *panchayat* was compelled to issue a license to the plant for three months in January 2006. But Coca-Cola began talking about shifting with compensation. Early 2006 also brought a new political party—the LDF—to power in Kerala. The LDF moved quickly to obtain reports on the pollution and soon banned the manufacture and sale of Coca-Cola and Pepsi-Cola in all of Kerala in August 2006.

Two key issues arise from the above discussion. The first is that the power and authority of the various state institutions in Kerala were not exercised in tandem. The lack of clear laws and rules on rights to water led to a protracted battle between state institutions and Coca-Cola. Although the devolution of power to local institutions to make decisions on resources was intended to enable the inclusion of citizens' voices, they often involve political parties and interests. In Plachimada, the *panchayat* is responsible for maintaining traditional drinking water sources, but it delayed its response to the problem of overexploitation and was then later restricted in terms of exercising its power. Determining the control of water resources must include the voices of the community to ensure social justice, particularly recognitional justice. The second issue is the complexity in understanding the role of the multitude of state institutions which all jockeyed to push their decisions, but also displayed ambivalence as to whether and when to support privatization.

The two above-mentioned issues are also intertwined in the case of the CCCS. As noted by Snitow and Kaufman (2008), "when private water companies case a city as a potential privatization target, one of the things they look for is a 'champion' in city government, someone who will take the lead in selling off the city's water services" (2008: 46). In Stockton, they found their champion in Mayor Gary Podesto, a former "big box" grocery store owner, who wanted to expand development, reinvigorate downtown, and streamline city services. The Mayor had an additional reason to privatize. Under pressure from federal and state environmental agencies to modernize sewage plants, cities like Stockton had to find ways to meet an expensive project. As such, the Mayor assumed that a private company could do it cheaply. But, as noted above, throughout the process, the Mayor declined to use the word "privatization."

The CCCS's concern was with the Mayor and the Council, but when the signature-gathering efforts fell shy by a few hundred signatures, Podesto was relieved and moved quickly to bring forward another deadline. Although there was a condition that no legal action would be taken up by either party, the CCCS lawsuit challenging the privatization had a court date in August 2003. In October 2003, the conservative Superior Court Judge Robert McNatt ruled that the city must unravel the deal in 180 days. He added that privatization itself creates a "significant change in the environmental status of the water department and requires an environmental impact report" (Snitow and Kaufman with Fox 2007: 45). He also called the city's self-declared exemption from such a report a misuse of discretion. The city appealed the ruling on a technicality and won as the court ordered a new trial sometime in the future. The issue remained in process for almost two years. Meanwhile, Podesto was forced to retire because of term limits. He decided to run for Senate, but lost the election.

Three years after the privatization, the legality of that action was unclear. In late 2006, the CCCS won another court decision that the privatization required environmental review. In early 2007, Stockton's new City Council voted not to appeal the court decision as it was dissatisfied with OMI/Thames' performance. The city and OMI/Thames arrived at an agreement to not criticize each other in public and set March 1, 2008 as the date on which the city would resume full control of its water system. Coalition activist Dale Stocking observed that this experience defies adoption of the private industry model over the public model. The victorious CCCS vowed to remain custodians of water as a public resource.

Privatization redefines ownership and control of resources that local people rely on for survival. Involved in this discourse is the NGO, an institution involved in the contentious politics at the village level; a politics that exacerbates differences based on gender, caste, and class. TBS as an NGO can be viewed as a resource for village residents, facilitating a dialogue amongst the residents about the governance mechanisms of the common property resource. Consider, for instance, the discussion of the effects of the drilling of borewells for those living at the lower end of the gradient, even though this benefited the richer farmers. Such discussions have been made possible by the formation of the "water parliament." While TBS is an established NGO, the water parliament session is for the village residents. The village residents are organized informally, but share a common sense of history in that they had collectively worked to revive the Arvari River, their main source of water.

Conclusion

This chapter is about the privatization agenda embedded in neoliberal policy that national governments are often compelled to adopt, as it is tied to the structural adjustment loan conditionalities put forth by the World Bank and the IMF. The local and global are intertwined with the state being the primary actor between those advocating policy (global) and the citizens/community (local). Thus, globally formulated policies, such as the neoliberal policies advocated by the World Bank and trade liberalization policies promoted by the WTO, impinge on the local, as is evident in the case of water.

Globalization processes which enable the neoliberal agenda have consequences for both developed and developing countries. As national, state, and local governments in countries such as Bolivia (Olivera 2004) and the United States (Snitow and Kaufman with Fox 2007), among many others, have enforced privatization in the name of efficiency, local citizen groups have organized to challenge the neoliberal tenet. The recognition of the threat to what has typically been common property to be owned and managed by local users has served as an impetus to organize challenges to state action. This has broader implications for recognizing the divergence between citizens' interests and state action, which is also complicated by the role of new institutional mechanisms such as NGOs at the local level.

The state and its policies continue to play a pivotal role in transposing and shaping neoliberal globalization at the national and local levels. The state may be implicated in the process of marketization and globalization, but it is constantly being shaped by the multiplicities of challenges and struggles of local groups. However, as water has been increasingly subjected to market imperatives and transformed into a means for capital accumulation, it has become increasingly difficult for governments to manage its distribution in ways that take into account social considerations (Swyngedouw 2005). While the emphasis on the local stems from the recognition of the need for decentralization, it is not necessarily a space of freedom or security because that space itself has its own hierarchy of power, with resources to defend the status quo and suppress dissent. Local social power based on gender, caste, and class precludes poor and small farmers, women, tribal groups such as the Adivasis, and racial minorities from participation in deliberations about control over water resources. The cases reinforce the need to recognize the power of local, village, town, or city-based groups as relevant to considerations of managing water resources collectively or by the public. All the cases discussed in this chapter resist the commodification of water by redefining water as a basic right, a social good that is a common resource of the people and not that of private corporations or state institutions.

Water itself is a local issue, not a national one. Making laws and policies for the control over and management of water does not imply that they are implemented and enforced as enforcement especially as the state is a sprawling apparatus with different institutions wielding power over different aspects of policies. Thus, the state enables and constrains the right to water. The state in between international agencies and forums on one side and community struggles on the other. This dynamic or ongoing dialogue with state institutions across the national and sub-national levels renders the state led neoliberal agenda ambivalent.

Overall, communities, advocacy groups, and activists around the world have been challenging and resisting privatization. In fact, community-based collective rights to water have also been raised with little success in global forums such as the World Water Forum (WWF). I will discuss these forums in the next chapter.

Notes

1. Plachimada is a small village in the Pallakad district of Kerala. It is located in the rain region of Western Ghats, and depends on groundwater and canal irrigation. The Perumatty *panchayat*, under which Plachimada falls, has 304 large tanks, of which 291 are private and 13 are public. Coca-Cola acquired 34.64 acres, mostly paddy fields, in the area in 1998 (Bijoy 2006).
2. But land distributed as of 2003 benefited only about 2000 Adivasi families of the almost 53,000 who were entitled to it (Bijoy 2006).
3. Rajasthan is a dry state with extreme temperatures during the summer.
4. The Coca-Cola Company operated in India after independence in 1947, but left the country in 1977 after a dispute with the Indian government regarding taxation and the right to shift its profits outside India. It returned in 1993, when the Indian government again encouraged foreign

investment and relaxed the regulations regarding foreign ownership of Indian companies. The basic idea was to allow foreign brands to compete with Indian brands, but within a few years the soft drinks market was completely dominated by two US-based firms: Coca-Cola and Pepsi (Berglund n.d.). The Hindustan Coca-Cola Beverages Private (HCCB) registered in 1993 as a subsidiary of Coca-Cola and invested more than US$1 billion between 1993 and 2003, established 27 wholly owned bottling operations, and a network of 29 contract-packers to manufacture a range of products (Bijoy 2006).
5. The *panchayat* is a local governance institution comprising elected members from the village.
6. In India, every state is divided into districts; districts comprise talukas, and several villages fall within a taluka.
7. The details of the case of the CCCS are drawn from three main sources: Snitow and Kaufman with Fox (2007); Snitow and Kaufman (2008); and Food & Water Watch (2007).
8. As per current exchange rates, $1 = 63 Rupees.

References

Agarwal, Anil. 2001. "Drought? Try Capturing the Rain." CSE Document.
Aiyer, Ananthakrishnan. 2007. "The Allure of the Transnational: Notes on Some Aspects of the Political Economy of Water in India." *Cultural Anthropology* 22(4): 640–658.
Bauer, C. J. 1997. "Bringing Water Markets Down to Earth: The Political Economy of Water Rights in Chile, 1976–95." *World Development* 25(5): 639–656.
Berglund, Henrik. n.d. Economic Globalization and Political Protest in India— The Case of Coca-Cola in Kerala. Unpublished paper.
Bijoy, C. R. 2006. "Kerala's Plachimada Struggle: A Narrative on Water and Governance Rights." *Economic and Political Weekly*, October 14.
Bijoy, C. R. 2008. "Forest Rights Struggle: The Adivasis Now Await a Settlement." *The American Behavioral Scientist* 51(12): 1755–1773.
Brewster, Marcia, Thora Martina Herrmann, Barbara Bleisch, and Rebecca Pearl. 2006. "A Gender Perspective on Water Resources and Sanitation." *Wagadu* 3: 1–23.
Briscoe, John, and R. P. S. Malik. 2006. *India's Water Economy: Bracing for a Turbulent Future*. The World Bank Agriculture and Rural Development Unit South Asia Region.
Chekutty, N. P. n.d. *Plachimada vs. Coca-Cola: 1000 Days On*. Accessed October 7, 2017. www.infochangeindia.org.

Food & Water Watch. 2007. *The Price of Privatization Stockton, CA*. Report.
Gies, Erica. 2009. "Is Water a Human Right or a Commodity? *World Watch* 22(2): 22–27.
Gleick, Peter H. 2010. *Bottled and Sold: The Story Behind Our Obsession with Bottled Water*. Washington, DC: Island Press.
Gujja, Biksham, K. J. Joy, Suhas Pranjape, Vinod Goud, and Shruti Vispute. 2008. *Water Conflicts in India: A Million Revolts in the Making*. London: Routledge.
Jayaraman, N. 2002. "No Water? Drink Coke". *CorpWatch India*. Accessed October 7, 2017. http://www.indiaresources.org/.
Jha, S. 2002. 'India's Water Wars', on *AlterNet*. Accessed October 7, 2017. http://www.mail-archive.com/bdcom@envirolink.org/msg05776.html.
Kaur, N. 2003. "Privatising Water." *Frontline* 20(18), April 30–September 12.
Koonan, Sujith. 2010. "Groundwater Legal Aspects of the Plachimada Dispute." In *Water Governance in Motion: Towards Socially and Environmentally Sustainable Water Laws*, edited by Philippe Cullet, Alix Gowlland-Gualtieri, Roopa Madhav, and Usha Ramanathanm, pp. 159–98. New Delhi: Cambridge University Press.
Krishnakumar, R. 2004a. "A Judicial Intervention." *Frontline* 21(2), January 17. Accessed October 7, 2017. http://www.flonnet.com/fl2102/stories/20040130003203400.htm.
———. 2004b. "Resistance in Kerala." *Frontline* 21(3), January 31. Accessed October 7, 2017. http://www.flonnet.com/fl2103/stories/20040213003703800.htm.
———. 2004c. "Kerala's Plight." *Frontline* 21(6), March 26.
Mahapatra R. 1999. "Waters of Life." *Down to Earth*, March 15.
Meinzen-Dick, Ruth S., Lynn R. Brown, Hilary Sims Feldstein, and Agnes R. Quisumbing. 1997. "Gender and Property Rights: Overview." *World Development* 25 (8): 1299–1302.
Menon, Ajit, Praveen Singh, Esha Shah, Sharachchandra Lélé, Suhas Paranjape, and K. J. Joy. 2007. "Chapter 4: Community-Based Natural Resource Management in Gopalpura, Rajasthan." In *Community-Based Natural Resource Management: Issues and Cases from South Asia*. New Delhi: Sage.
Narasimhan, T. N., and Vinod K. Gaur. 2009. *A Framework for India's Water Policy*. Bangalore: National Institute of Advanced Studies.
News Feature. 2007. "A Breath of Fresh Air." *Nature* 445 (15): 706–708.
News Hour. 2008. "Coca-Cola Plants Put Strain on Water Resources in India." The Online News Hour with Jim Lehrer (PBS).
Olivera, O. 2004. *Cochabamba! Water War in Bolivia*. Cambridge, MA: South End Press.
Pillai, P. R. Sreemahadevan. 2008. *The Saga of Plachimada*. Mumbai: Vikas Adhyayan Kendra Publications.

Priscoli, J. D., J. C. I. Dooge, and R. Llamas. 2004. *Water and Ethics: Overview.* Paris: UNESCO.
Raman, K. Ravi. 2005. "Corporate Violence, Legal Nuances and Political Ecology: Cola War in Plachimada." *Economic and Political Weekly* 40(25), June 18: 2481–2486.
Raman, K. Ravi. 2007. "Community-Coca-Cola Interface: Political-Anthropological Concerns on Corporate Social Responsibility." *Social Analysis* 51(3): 103–120.
Raman, K. Ravi. 2010. "Transverse Solidarity: Water, Power, and Resistance." *Review of Radical Political Economics* 42(2): 251–268.
Sankaran, K. 2009. "Water in India—Constitutional Perspectives." In *Water and the Laws of India*, edited by Ramaswamy R. Iyer, pp. 17–31. New Delhi: Sage.
Shiva, Vandana. 2005. *Earth Democracy: Justice, Sustainability, and Peace.* Cambridge, MA: Southend Press.
Shree, Ruchi. n.d. "Plachimada Against Coke: People's Struggle for Water." Accessed October 7, 2017. http://conflicts.indiawaterportal.org/sites/conflicts.indiawaterportal.org/files/Ruchi%20Shree-%20Plachimada%20against%20Coke-sh.pdf.
Sinha, Dipa. 2016. *Women, Health and Public Services in India: Why Are States Different?* New York: Routledge.
Snitow A, D. Kaufman, and M. Fox. 2007. *Thirst: Fighting the Corporate Theft of Our Water.* San Francisco, CA: Jossey-Bass.
Snitow, Alan, and Deborah Kaufman. 2008. "The New Corporate Threat to Our Water." In *Water Consciousness: How We All Have to Change to Protect Our Most Critical Resources*, edited by Tara Lohan, pp. 45–57. San Francisco, CA: AlterNet Books; Healdsburg, CA: Watershed Media.
Subramaniam, Mangala. 2014. "Neoliberalism and Water Rights: Case of India." *Current Sociology* 62(3): 393–411.
Swyngedouw, Erik. 2005. Dispossessing H_2O: The Contested Terrain of Water Privatization. *Capitalism Nature Socialism* 16(1): 81–98.
Tarun Bharat Sangh. 2006. *Arvari Sansad.* Jaipur: Jal Biradari.
United Nations. 2005. "Women and Water." *Women 2000 and Beyond.* New York: Division for the Advancement of Women, Department of Economic and Social Affairs of the United Nations Secretariat.

CHAPTER 4

Controlling Water Resources from "Above": Global Water Forums

The issues of ownership and management of water have been key to the localized struggles of communities and citizens across the developed and developing world as discussed in Chapter 3. Many of these struggles are against the privatization efforts of large multinational corporations. As alluded to in Chapter 3, major private companies have entered the American market: Suez and Veolia from France and Germany's utility corporation RWE/Thames. These three corporations have dominated the global water business and are among the world's largest companies. Together they control subsidiaries in more than 100 countries.

In fact, who controls water is also at the center of discussions in global forums such as the World Water Forum (WWF), except that they make little reference to the right to water. The World Water Council (WWC), which organizes the WWF once every three years, is an international think tank based in Marseille, France. Its goals are to raise awareness of global water issues. The Council is stated to catalyze "collective action in between each WWF and guarantees continuity and success from one edition to another" (World Water Council Report 2014). The WWC has direct links to two of the world's largest water corporations, Suez and Veolia. The President of the WWC is also president of a company equally owned by Veolia and a subsidiary of Suez. Among the WWC members are the World Bank and other financial institutions which promote water privatization. A WWC report notes that the business industry makes up 41% of the membership of the WWC. The WWC membership sets the WWF's agenda.[1]

© The Author(s) 2018
M. Subramaniam, *Contesting Water Rights*,
https://doi.org/10.1007/978-3-319-74627-2_4

Critics of the WWF draw attention to the goals of the WWC members who came together to create the WWC in 1996. The WWC promotes public–private partnerships that have put water services in some Latin American countries under private ownership. An alternative to the WWF is the People's Water Forum (PWF), which represents the rural poor, the environment, and organized labor pushing for the elimination of water privatization (www.peopleswaterforum.org). The PWF was soon renamed the Alternative World Water Forum (AWWF) in Florence, Italy in 2003. The organization committee of the AWWF consists of 24 "international" organizations. In addition, 32 French organizations and 41 local organizations in Marseille for a total of 97 organizations that put together the 2012 AWWF. Over 150 organizations and 384 individuals also signed the "declaration of participants."

The Global Water Justice Movement

The global water justice movement comprises a variety of groups/organizations and activists across countries. Grassroots action like the cases discussed in Chapter 3 are also connected to a broader global movement. The various groups and organizations have been mobilizing their communities for constitutional recognition of the right to water. The global water movement, in its many forms, has given birth to national water rights networks (such as in Italy) and continental networks (in Africa), and has led to the decline of water privatization in Latin America, Africa, Europe, and notably in France. Although France is home to the main multinational water companies, in Paris itself, water resources are publicly managed. The water movement has participated in the fight to have access to water recognized as a fundamental human right. The movement has particularly sought the attention of the United Nations (UN) because a UN Covenant would set the framework for water as a social and cultural asset, not an economic commodity.[2] Moreover, it would establish that private control of water is incompatible with water as a "commons." In other words, neither the state nor international institutions should support interventions that entail the privatization of water resources and exclude those who have equal access to water.

Water was not included in the 1947 UN Universal Declaration of Human Rights because at that time water was not perceived to have a human rights dimension. Because water as a human right is not entirely enforceable, policy making about water has shifted from the UN and even governments to institutions and organizations, such as the World Bank and the WWC, which support the privatization and commodification of water. However, for over a decade, there have been calls for a right to water convention (Barlow 2008). Local groups and NGOs were increasingly concerned about water companies going global and most such companies were backed by global financial institutions. International laws were viewed as being essential to address such concerns. The 1990 Rio Earth Summit had noted several key areas of concern, which included water, climate change, biodiversity and desertification. All of these issues have resulted in UN conventions, except for water.

The intense lobbying by various civil society groups has led to the recognition of water rights in several UN resolutions and declarations (Barlow 2008; see Appendix 4.1 for a list of resolutions). The Habitat Agenda (1996), the Millennium Summit (2000), and the World Summit on Sustainable Development (2002) have underscored the importance of water for development and have set goals and targets (Third World Water Forum Analysis 2004). In addition, there are the 2000 General Assembly Resolution on the Right to Development, the 2004 Committee on Human Rights Resolution on toxic waste and the May 2005 statement by the 116-member Non-Aligned Movement on the right to water for all.

The UN General Assembly agreed in 2000 on the Millennium Development Goals (MDGs), a set of eight goals for development and poverty eradication by 2015. Water-related targets were to halve the proportion of people without access to sustainable drinking water (United Nations 2002) and to halve the proportion of people who are hungry. The MDGs appeared to be a very powerful instrument for more positive and result-oriented international discussions and development of policies. Most important is General Comment Number 15, adopted in 2002 by the UN Committee on Economic, Social and Cultural Rights, which recognized that the right to water is a prerequisite for realizing all other human rights.[3] However, General Comment Number 15 is an interpretation and not a binding treaty or convention (Scanlon et al. 2004). This is also discernable from the clarifications in an undated UN document made available to the media (UN Water Decade Programme on Advocacy and Communication and Water Supply and Sanitation Collaborative Council n.d.). Noted as clarifications of common misconceptions, the brief asserts that the right is not about the free or unlimited use of water, but that it should be affordable for all and that people are expected to contribute financially or otherwise to the extent that they can do so. In addition, the state is noted to be the institution that takes steps to progressively realize this right. The lack of clarity in the role of the community—local residents/citizens—to manage and control water resources has led activists to develop formal networks.

In early 2004, Danuta Sacher of Germany's *Bread for the World* and Ashfaq Khalfan of the *Right to Water Program* at the UN Centre on Housing Rights and Evictions called a summit that resulted in a new international network called Friends of the Right to Water. The network began to mobilize other water justice groups and national governments to join the campaign to strengthen the rights established in General Comment Number 15 and put in place the mechanisms to ensure the implementation of the right to water through a covenant. By 2006, there was some progress toward the consideration of the issues of access to water in international human rights instruments. While it did not result in a specific covenant, there was some optimism about the potential of the incorporation of the right to water in international instruments. In fact, in April 2007, Anil Naidoo of the Council of Canadians' Blue Planet Project (also a founding member of Friends of the Right to Water) organized the presentation of a letter of endorsement calling for a right to water covenant, signed by 176 groups from across the world, to UN Commissioner Arbour (Barlow 2008).

Specifically, it has been argued that the adoption of the human right to water (HRW) in international discourses and institutions marks significant progress toward redressing inequities (Perera 2014; Mirosa and Harris 2012; Sultana and Loftus 2012). Yet, a number of criticisms have also been put forth. In a major critique, Bakker (2007) questions the degree to which HRW serves the anti-privatization agenda, given that there is no reason why private companies cannot be part and parcel of an HRW agenda, considering that many of them (private companies) have adopted HRW as a discourse. In a similar vein, Parmar (2008) cites evidence that the HRW can be particularly hostile to the needs and goals of historically marginalized communities, including indigenous populations, given the Western bias on the "rights" framework that is often about individual rights. Mehta (2006) problematizes the HRW focus on domestic water rather than livelihood needs, and also highlights problems with implementation that can leave marginalized communities with sub-standard access. Further critiques show how technical framings of the HRW do not allow much room for social mediation or consideration of the on-the-ground realities and lived experiences of inequitable water access (Goff and Crow 2014).

These arguments demonstrate that control over and rights to water may be worsened under the guise of HRW. This risk is particularly salient when implementation is presented as, or is assumed to be, just and impartial. Consider, for instance, the case of South Africa, where constitutional guarantees of the HRW are in place (Scanlon et al. 2004), and yet evidence shows ongoing difficulties and inequalities in securing affordable or safe access to water, especially for the most impoverished and marginalized people (Loftus 2006; Desai 2002).

Yet, the passing of the 2010 UN Resolution, as well as its adoption in a range of national constitutions, suggests that the HRW as a discourse and policy will remain. A number of water justice activists also view the HRW as a potentially effective mechanism in making progress toward the goal of safe and affordable water for all regardless of their ability to pay (Mirosa and Harris 2012; Barlow 2008). However, it is important to recognize that there is an urgent need for dialogue on the meaning and implementation of the HRW, and particularly the role and obligations of the state in adopting international covenants rather than "hiding behind capacity arguments" (Barlow 2008: 180).

Despite the positive developments in the global arena, there is some concern that the efforts to move the discussions about water rights will be hijacked by the water corporations and institutions such as the World Bank. Until recently, the World Bank and private corporations forcefully opposed a right to water convention, as did many European countries such as France, England, and Germany, which are home to the big water companies. As international agencies such as the UN have engaged in and articulated the meaning of the right to water and global networks have emerged to demand clarity on rights of the people to water, it is essential to discuss the main global forums at the center of these issues. I begin with a discussion of the WWF—its purpose, its membership and structure, and the attendance and agenda at each Forum—as explained in the documents that have regularly been released by the WWC (see the list of documents in Appendix 4.2).

THE WWF AND THE WWC

The WWF is a triennial global summit organized primarily by the WWC. Recognizing a vacuum in governance concerning water policy, the idea of a WWC was originally proposed in 1992 during a meeting at the UN's International Conference on Environment and

Development in Dublin and at the Rio de Janeiro Earth Summit. In 1994, at a special session of the 8th World Water Congress in Cairo organized by the International Water Resources Association (IWRA), a resolution was passed to create the WWC. The WWC was officially incorporated in 1996 in Marseille.[4] It had three official founding members who drafted its constitution and by-laws; Dr. Mahmoud Abu-Zeid of the Egyptian Ministry of Public Works and Water Resources, Aly Shady of the Canadian International Development Agency, and René Coulomb of Suez Lyonnaise des Eaux, a French private water corporation (WWC Constitution & By-Laws 2013–2015). The WWF is described as a key factor in a "collective response—as a global water community—to addressing these challenges and finding solutions that work for the benefit of all" (World Water Forum 2009: 5).

The Purpose of the WWC and the WWF

The WWC's stated mission is to "promote awareness, build political commitment and trigger action on critical water issues at all levels, including the highest decision-making level, to facilitate the efficient conservation, protection, development, planning, management and use of water in all its dimensions on an environmentally sustainable basis for the benefit of all life on earth."[5] However, "authority over the world's water is fragmented among the nations of the world, hundreds of thousands of local governments, and countless non-governmental and private organizations as well as a large number of international bodies" (WWC Constitution & By-Laws 2013–2015: 4). Therefore, the WWC exists as an umbrella organization designed to network together intergovernmental, governmental, private industry, professional, academic, and civil society organizations. According to its website, the WWC describes itself as a "membership organisation representing and federating the great diversity of water stakeholders."[6] Through its primary flagship, the WWF, it seeks to "encourage and strengthen multi-stakeholder, multi-disciplinary and cross-sectorial exchange and dialogue" in order to establish a global consensus on the effective management of the world's water resources.

Membership and Structure of the WWC

Annual membership fees to the WWC range from €300 to €3000. The WWC currently consists of over 300 member organizations, such as the World Bank and UN agencies (the United Nations Development

Programme (UNDP) and the United Nations Educational, Scientific and Cultural Organization (UNESCO)). Well over half of its members are from the private sector, including private water, power, construction, engineering, and manufacturing companies and professional organizations (WWC List of Members 2014). Each member organization, via its appointed representative, gets one vote to elect each of the 35 members of the Board of Governors (36 including the city of Marseille), which is decided by a simple majority. The Board in turn elects the six-member Bureau, including the president, vice-president, and treasurer, who make the majority of ruling decisions. The current President is Benedito Braga, a civil engineer from the Polytechnic School of the University of Sao Paulo, Brazil, and Vice-Chairperson for Latin America and the Caribbean region of the International Hydrological Programme (IHP) of UNESCO. The Vice-President is Dogan Altinbilek, who represents the Turkish Contractor Association. The WWC has been criticized for being overly representative of private industry, particularly during the tenure of its previous President, Loïc Fauchon (Barlow and Clarke 2002). He was also the President of Société des Eaux de Marseille (SEM), the main company of the Groupe des Eaux de Marseille, the fourth-largest French group in the private water sector, since 1991. The Groupe des Eaux de Marseille is jointly owned by Veolia and is a subsidiary of Suez, two of the largest private water corporations in the world.

The WWF: Attendance and Agendas

The WWF is held every three years for the purpose of directing the policy and management of the world's water resources. The first was held in Marrakesh in 1997, followed by The Hague in 2000, Kyoto in 2003, Mexico City in 2006, Istanbul in 2009, Marseille in 2012, and Daegu and Gyeongbuk, Korea in 2015. The 2018 WWF is planned to take place in Brazil (as at the time of writing this book).

The first WWF in Marrakesh was fairly small, consisting of four sessions and 36 presenters almost exclusively from governmental, intergovernmental, and private corporate organizations focusing mainly on water management, development, and globalization.[7] It sought to establish the voice and leadership of the WWC in water affairs through a document called the Marrakesh Declaration, and called on the WWC to develop a "Long Term Vision for Water, Life, and Environment in for the 21st Century," or World Water Vision for short (Table 4.1).

Table 4.1 World Water Forums

Year	Location	Key focus
1997	Marrakesh, Morocco	Focused on water management, development, and globalization. Aimed to establish the voice and leadership of WWC in water affairs
2000	The Hague, The Netherlands	Released World Water Vision. Noted that demand for clean water is outpacing supply and emerging crisis should be stemmed by 2025
2003	Kyoto, Japan	Focused on World Water Action Report, which documented more than 3000 local (village to national-level) projects
2006	Mexico City, Mexico	Focused on local actions for a global challenge. Right to water was debated for the first time
2009	Istanbul, Turkey	Focused on bridging divides for water. Local authorities encouraged to engage in local projects for adopting and updating water infrastructure
2012	Marseille, France	Officially incorporated language of "water rights." Did not provide details of the meaning of 'rights'
2015	Daegu and Gyeongbuk, Korea	Focused on "Water for Our Future." Stands for willingness to move forward in the future with concrete actions and implementation of solutions to water challenges. Citizens' Forum was organized

The World Water Vision was presented at the second WWF at The Hague, in 2000. The message of the Vision is particularly aimed at "the leaders and professionals who have the power and knowledge to help people turn visions into reality" (WWC World Water Vision Report 2000: xiii). It states that the demand for clean, safe water is outpacing supply and that the current course of action will ultimately lead to crisis. However, they believe the crisis can be stemmed by the year 2025 through "international cooperation, integrated management, public funding in research and innovation, full-cost pricing of all water services, and massive increases in water investments" (WWC World Water Vision Report 2000: xxvi).

The second WWF consisted of approximately 5700 international participants discussing and debating a wide variety of local and global water-related issues, including 113 ministers and officials from 130 countries who participated in the Ministerial Conference. This resulted in the *Ministerial Declaration of The Hague on Water Security in the 21st Century* (2000), which states that water is vital for people and ecosystems as well as the development of countries. While the statement emphasizes bringing together stakeholders at all levels, the emphasis is mainly on efforts at the national and sub-national levels, and references to management and investments. Even as they say it is the poor who are most in need of access to safe water, they also say it needs to be at an affordable cost. Following on from this theme, according to the WWC website, in 2001 the WWC put the issue of financing at the top of its priorities by establishing the Panel of Financing Water Infrastructure.

The third WWF was held in Kyoto, Siga, and Osaka, Japan in 2003 and was the largest water conference up to that point. It consisted of over 24,000 participants from 183 countries from all regions of the world, although Japanese participants outnumbered all other regions by at least 18–1 (Third WWF Final Report 2003: 7). Participants were organized into 351 sessions, 33 themes, and five regions. According to the official report, there were also 130 ministers from 103 countries in attendance. Among the EXPO exhibitions, there were approximately 345 private companies, along with a few dozen universities, NGOs, and municipal organizations. The vast majority of these were from Japan.

The focus of the conference was the World Water Action Report, which documented more than 3000 local (from village to national-level) actions related to the World Water Vision, including projects, applied research and studies, awareness-raising campaigns, and policy, legal, and institutional reforms. The sessions themselves also focused on commitment to action; each session organizer was required to produce a statement of concrete action resulting from their respective session, and more than 100 new initiatives for improved water management were launched. While the themes of the session covered a wide variety of topics, including water as a human right and gender equality in water-related decision-making, the voice of industry had the most impressive resonance. A few sessions included indigenous peoples and activist groups, but most focused on managing water as a strategic resource, industrial development, privatization, financing, and scientific research.

The fourth WWF was held in Mexico City in 2006. It was slightly smaller than the previous one, consisting of about 20,000 people plus about 5000 organizations from 168 countries, including governmental delegations from 148 countries, 200 legislators, and 160 representatives of local authorities. There was again an EXPO of about

340 private companies, along with a Water Fair consisting of about 100 governmental, NGO, and regional organizations, who shared an open space. A total of 73% of participants were from Mexico (Fourth WWF Final Report 2006: 24). This Forum aimed to bring together politicians, water management professionals, and NGOs to aid cooperation for "better living standards for people all over the world and a more responsible social behavior towards water issues in line with the pursuit of sustainable development."[8]

The fourth WWF was organized into 206 sessions under the theme "local actions for a global challenge." A total of 12% of the sessions focused on specific regions, while 88% had a global focus. "Local action" really meant big infrastructure projects in local places like Nepal and Morocco. According to the synthesis report, discussions about infrastructure dominated the conference, and contrasted with the conservation approaches of indigenous and local peoples. North American and Asian development banks announced they would be loaning hundreds of millions of dollars to water infrastructure projects in the coming years (Fourth WWF Final Report 2006: 7). Interestingly, the report also states that the right to water was for "the first time debated in a rather consensual spirit" (Synthesis of the Fourth World Water Forum 2006: 2). The end of the report also noted the political nature of the water debate, as both a basic right and an economic commodity at the same time. This was particularly apparent in the Ministerial Conference, in which the Right to Water was addressed. "There are still contrasting views on this issue, but the various political declarations show that the concept is not merely ideological, but that it has a concrete meaning for all and is endorsed by many, starting with the local authorities and the parliamentarians" (Synthesis of the Fourth World Water Forum 2006: 108).

According to the India Resource Center, Coca-Cola's sponsorship of the WWF, the global summit aiming to improve access to freshwater worldwide, is out of line with the event's stated objectives. This sponsorship is described as a publicity maneuver to gloss over its record of wasteful water use and pollution that people in various countries have been protesting against (see Chapter 3 for the case of Plachimada and below for the broader global perspective).

The fifth WWF was held in Istanbul, Turkey, in 2009. This took place just as Turkey was planning one of the most sweeping water privatization programs in the world. There were approximately 16,000 engaged participants from 182 countries, including 1300 political process participants, 165 delegations, 90 ministers, and 19 undersecretaries, over 250 parliamentarians, and over 300 mayors and local officials (Fifth WWF Highlights from Istanbul 2009: 2). The major theme was "bridging divides for water" and the areas of focus included risk management, development, security, governance, finance, and education. The Istanbul Water Consensus encouraged local authorities to engage in local projects by adapting and updating their water infrastructure, and gained over 1000 signatures. The final report stated ironically that: "At the dawn of a new era, the fifth World Water Forum marked the rise of a new paradigm, a turning from the production-oriented to the eco-oriented" (WWC Water at a Crossroads 2009: 7).

The sixth WWF in Marseille in 2012 officially incorporated the language of "water rights" into its 2012 Ministerial Declaration, but without any details on its meaning (see the discussion in the next section). The Right to Water appears throughout the final report and in the conference sessions, although themes of global water governance and security each came before in the 10 high levels panels. The sixth WWF included about 30,000 participants. Fifteen heads of state, government, UN agencies and European commissioners, and 112 ministers, vice-ministers, and secretaries of state from 145 countries participated in the Ministerial Conference. More than 350 local and regional officials were also present, along with 250 parliamentarians. More than 3500 NGO and civil society representatives were present. At the EXPO, there were pavilions for over 40 countries as well as local and regional authorities, over 60 public and private enterprises, facilities, research and academic institutes, and close to 40 intergovernmental organizations, international institutions, NGOs, civil society, and not-for-profit associations.

Following the sixth WWF, WWC President Loïc Fauchon was replaced by Benedito Braga in 2013. This represented a big image change for the WWC and before the seventh WWF. The slogan of the seventh WWF was "Water for Our Future" (Seventh World Water Forum 2nd Announcement 2014). The seventh WWF was jointly organized by the Republic of Korea and the WWC, together with Daegu Metropolitan City and Gyeongbuk Province. It stands for the willingness to move forward to the future with concrete actions and the implementation of solutions to water challenges. Over 40,000 participants from water-related government agencies, international organizations, NGOs, academia, corporations, and the media representing 168 countries attended approximately 400 sessions, making this WWF the largest in the Forum's history

(World Water Council 2015). About 300 suggestions were derived from sessions organized as thematic process; political process, regional process, science and technology process; and citizens forum. In addition, more than 420 organizations supported the various working groups and sessions (Seventh World Water Forum 2nd Announcement 2014).

The WWF and Water Justice: Water Is a Human Need and (?) a Human Right

As the global water justice movement gained momentum, private entities and global institutions joined the dialogues about water rights. Civil society groups, as discussed above, feared that their efforts would be co-opted by corporations. The second WWF asserted a common goal of providing water security in the twenty-first century, which the Ministerial Declaration noted was "reflected in the unprecedented process of broad participation and discussion by experts, stakeholders and government officials in many regions of the world" (World Water Forum 2000: 1). Moreover, the process had benefited from the "World Water Vision" launched at the first WWF in Marrakesh, from the formation of the World Commission on Water in the twenty-first Century and from the development of the Framework for Action by the Global Water Partnership. While the main challenges identified included meeting basic needs and governing water wisely, among others, the list of actions focused on a wide variety of aspects that mentioned stakeholders, the UN, and other multilateral institutions. Despite a reference to "open, participatory and transparent" follow-up actions, there is no mention of citizens and communities directly seeking rights to water, or of state accountability to provide water services. At the second and third WWFs held at The Hague and Kyoto, respectively, WWC members and governments rejected civil society calls for a right-to-water convention and said that water is a human need, not a human right. As Barlow notes, "you cannot trade or sell a human right or deny it to someone on the basis of inability to pay" (2008: 181).

An analysis document of the third WWF at Kyoto notes that water for basic needs is officially a human right since the UN Economic and Social Council agreed to this in 2002. But it points out that as long as "countries do not integrate this into their constitutions, their legal frameworks, and their policies, it does not affect or benefit people, especially the poorest" (Third WWF Final Report 2003: 20). Many participants at the third WWF were frustrated that no agreement could be reached to incorporate that statement into the Ministerial Declaration (Third WWF Final Report 2003). They suggested that the "water community should try to make proposals to the policy makers on how to implement the principle" (Third WWF Final Report 2003: 20). In doing so, the WWF was somewhat absolving itself of the key needs and rights of the water community to which it made reference.

At the fourth WWF held at Mexico City, the ministerial declaration once again did not include the right to water. But the WWC did release a new report entitled *The Right to Water: From Concept to Implementation*. This report restates the various UN documents (many of them referred to above) without any reference to privatization. It notes that the right to water can be accomplished in a variety of ways without details. While there is no discussion of the public–private divide about water rights, the foreword of the report written by Loïc Fauchon, the WWC President of the Sixth WWF, links water to human dignity, an aspect that civil society groups also emphasize.

The fifth WWF held at Istanbul focused on the need for the international water community to work together in order to bring about the "bridging [of] divides for water" (Fifth WWF Highlights from Istanbul 2009). The highlights of the Istanbul WWF reiterate the point made during the third WWF about the inclusion of the right to water in the Ministerial Statement. "Despite the difficulty for a very small number of countries to agree in principle to the inclusion of the Right to Water in the Ministerial Statement, 32 countries have already" (Fifth WWF Highlights from Istanbul 2009: 2). It observed that the main debate concerned the distinction between water as a human right or as a social and economic right, the latter concept already being widely accepted. The Forum therefore looked to the UN Committee on Human Rights for a thorough evaluation of the implications of water as a human right. Interestingly, the parliamentarians at the fifth WWF requested the WWC to establish a "permanent international Parliamentary 'helpdesk' to aid political cooperation on water legislation and its implementation" (Fifth WWF Highlights from Istanbul 2009: 3). The "helpdesk" was to undertake four key types of action: to coordinate by guiding responsibility for enacting water and sanitation laws; to showcase best global practices and water-related legislation; to respond by quickly answering urgent questions; and to link parties in a globally cooperative interparliamentarian network.

As in prior WWFs, several key issues were covered at Istanbul, including the need for public participation for the good governance of water resources. Such assertions have not been translated into practice despite almost two decades of meetings of the Forums. Although the fifth WWF's official program heralded "more diverse participation mechanisms," the WWC refused to allow the President of the UN General Assembly a public audience. President Miguel d'Escoto has been an outspoken critic of water privatization. The fifth WWF also held the first "Heads of State" meeting in which major national leaders debated the Right to Water. In the official Ministerial Statement, a declaration of the Right to Water was rejected. Instead, it was framed as a "basic need" in the final nonbinding communiqué. Spain and Latin American countries were notable dissenters, but they were outweighed by the United States and China. Despite the fifth WWF declaration, in 2010 the UN General Assembly and Human Rights Council issued resolutions declaring a HRW.

The sixth WWF differed somewhat from earlier WWFs in three main ways (World Water Forum 2012). The first is that the Forum's theme of "access to water for all" was aimed to use the cash surplus—over €1 million—remaining after the liquidation of the International Forum Committee to support projects on the ground promoting access to water and sanitation for all. Sixteen projects across more than a dozen countries and intended to reach nearly 100,000 people were funded by the WWC and were evaluated. Second, the WWC acknowledged the indispensable role of local stakeholders and grassroots organizations in addressing water-related challenges, as well as their capacity-building water efforts and services aimed at improving the living standards of the communities. This is a major step in the goals and aims of the WWC and the WWF. Third, in a step beyond earlier assertions, WWC noted that "water for all will only be possible by connecting local realities to global matters" (World Water Forum 2000: Foreword). These key issues seem to suggest that the gap between the WWC and the AWWF is narrowing but this is not entirely simple because they do not reflect open dialogues between citizens and those who represent corporate interests. In an effort to enable the participation of citizens, the seventh WWF in Daegu and Gyeongbuk, Korea included a Citizen's Forum.

The Citizen's Forum was intended as a mechanism to reach out to citizens by various means. Programs and activities within the Forum were to focus on specific stakeholder groups such as women, children, and youth (Seventh World Water Forum 2nd Announcement 2014).

The announcement urged civil society stakeholders to engage in the discussions. It was the first opportunity for civil society groups to meet and share their expectations for the seventh WWF. For the organized preparation of the seventh WWF, the International Steering Committee (ISC) was established jointly by the National Committee of Korea and the WWC in October 2012. This committee included the Jal Bhagirathi Foundation (JBF), India. JBF was established as a nonprofit organization on 15 January 2002 in response to the burgeoning water crises facing the Thar Desert area in the state of Rajasthan, western India, and the vast potential for participatory water management as a path to water security. Rajendra Singh, the "waterman" discussed in Chapter 3, is a founder trustee of the JBF.

The final report of the seventh WWF reiterates that access to water and sanitation was recognized as a human right in 2010. "Implementing this fundamental right is a legally binding obligation. This right entitles each person to have sufficient supply of water for personal and domestic uses, access to water of safe quality, being accessible physically and continuously, culturally acceptable and affordable without discrimination" (World Water Council 2015: 33–34). In a discussion of this issue, there were presentations of concrete examples of how the provision of water services can be improved and how stakeholders' engagement can facilitate the implementation of the right to water. Issues relating to women and water were prominently discussed in the Citizens' Forum. In fact, "Gender Equality to Achieve Future Water Security" was one of the 31 sessions of the Forum. However, there is a lack of clarity on how these issues will be incorporated into the making and implementing of policies that includes local citizens who seek a right to water.

The year 2015 holds a special meaning in terms of formulating actions for the global water agenda. As noted in the final report of the seventh WWF, it was the target year of the MDG, which was to be followed by a new global sustainable development agenda to be confirmed at the UN Summit on Sustainable Development in September (Synthesis Report Seventh World Water Forum 2015). Greater political commitment was therefore expected from national governments. The Ministerial Declaration focused on plans and actions to advance water-related cooperation at a global scale in order to ensure progressive access to water and sanitation for all. In addition, the ministers reinforced the need for a dedicated water goal in the post-2015 Development Agenda.

The agenda of the WWC and WWF has been revised since the first one in 1997. There has been some reference to water rights, action for change, and the urgent need for a global response to a growing crisis. But it is unclear if there any effort has been made to untangle private corporate interests and the pursuit of profits from community struggles for control over and seeking access to water resources. The WWC is working with Green Cross International, an environmental organization headed by Mikhail Gorbachev, the former President of the Soviet Union. This organization has launched a very high-profile campaign for a UN convention on the right to water. While Green Cross International in its draft convention admits to excessive profits in water exploitation by private corporations, it also equalizes the right to water with private financing for water services and private management of water utilities (as in the Stockton case discussed in Chapter 3), and calls for market systems to follow market rules. Canada's Blue Planet project has critiqued the approach outlined in the draft convention and argues that it does not represent the global movement's demand for a right to water. Yet Gorbachev has defended the pro-corporate proposal, noting that private corporations alone have the intellectual and financial potential to solve the global problems relating to water (Barlow 2008).

The global movement for water justice is unlikely to accept the draft convention of Green Cross International because water is a human right and a public resource. Moreover, the UN Covenant will have to address two limitations of existing human rights law to make it acceptable to civil society groups. These two limitations are the failure to establish relevant enforcement mechanisms and the inability to bind international bodies. In other words, these rules reinforce the notion of water as an economic good that prevents national governments and other state institutions from making and maintaining laws and practices needed to protect human rights which would inhibit the private rights ingrained in these agreements. It enables corporations with powerful instruments to affirm their rights over water, which states cannot refuse. This has also led to difficulties for states in terms of addressing conflicts between human rights norms and trade laws. So unless the UN asserts its framework, the primacy of the HRW, it will become a bystander in the ongoing conflicts between communities/citizens' rights to water, and private entities and the state. In this context, it is interesting to trace the emergence of the AWWF and its goals.

THE AWWF

As an alternative to the WWF, activists and scholars representing the rural poor, the environment, and organized labor organized what was initially called the PWF. For several years, different civil society movements have fought side by side for water conservation and management of water by citizens. Activists have created platforms, propositions, and campaigns at events such as the 2003 AWWF in Florence, Italy and subsequent Forums, as well as within international Social Forums such as those in Porto Alegre, Caracas, Nairobi, and Belem. These gatherings helped solidify the movement to reappropriate water, a communal resource which belongs to all of humanity.

It is important to note that most civil society groups and activists demanding water rights, including the control over and management of water resources, initially participated in the WWF. Evidence of this is available in the documentary *Thirst*. This participation often involved questioning the water-as-a-commodity approach emphasized by the WWF. As noted above, at the second and third WWFs held at The Hague and Kyoto, respectively, WWC members and governments declined to consider civil society calls for a right-to-water convention and said that water is a human need, not a human right. Some NGOs have continued to attend the WWC; for example, the JBF attended the seventh WWF.

The WWF is not open to the public and fees of hundreds of euros are required for participation. For instance, the fee for attending the seventh WWF ranged from €155 for a one-day pass to €520 for a Forum pass (that is, for the entire week). Attendance also requires special security screenings. As a result, many local citizens in the host cities are excluded from taking part in the deliberations of the WWF. Large protests have been organized at most WWFs. In Mexico City in 2006, an estimated 10,000–20,000 people marched in protest against water privatization, which is nearly as many people as were in the WWF. According to the Freshwater Action Network, high levels of security and the cost of entry prohibited many local people and Mexican groups from attending. This led to further accusations that the Forum was an undemocratic, exclusive talking-shop. However, on a more positive note, most sessions did seek to explore pro-poor and sustainable solutions to water management, and were of a high standard and well attended.[9] The 2009 WWF saw some of the biggest protests in Turkey and peaceful protesters were shot with water cannons and gas by the police. Several peaceful protesters were deported by the Turkish government.[10] Extensive protests also took place in Marseille in 2012.[11]

The first PWF was held in Florence in 2003 and attracted more than 1400 participants (70% of them Italian). There have also been Alternative World Water Forums (AWWF) in 2005 in Geneva, in 2006 in Mexico City, in 2009 in Istanbul, in 2012 in Marseille, and in 2015 in Daegu and Gyeongbuk, Korea. Each of these Forums is organized by different local coalitions. As a result, information about the AWWFs is sparse, expect for some details about the 2009 and 2012 Forums. The lack of details also speaks to the lack of resources among civil society groups and therefore their inability to develop glossy action-oriented reports that can be disseminated to the public (see further below) (Table 4.2).

The AWWF has declared that it will continue research to find solutions to the worldwide water crisis and make the "water movement structure sustainable." The AWWF has online space for sharing information and details about water-related protests called "water voices." The AWWF is pluralistic, diverse, nonreligious, nongovernmental, and is not tied to any political party. It expresses positions in a decentralized manner via networks and undertakes concrete actions at both the local and international levels. The aim is to find alternative ways to protect water resources and manage them in an ecological, public, and

Table 4.2 Alternative World Water Forums

Year	Location	Key focus
2003	Florence, Italy	Started as People's Water Forum (PWF) to assert right to water
2005	Geneva, Switzerland	No information available
2006	Mexico City, Mexico	Rejected all forms of privatization. Called for solidarity between present and future generations
2009	Istanbul, Turkey	Recognized interdependence between water and climate change. Determined to defend water as a common good
2012	Marseille, France	Asserted that WWC speaks for transnational companies and the World Bank. Charter of Action was released
2015	Daegu and Gyeongbuk, Korea	Declared a commitment to strong public water services for all. Declared that WWF is illegitimate and that water and sanitation are a human right

participatory manner. It will always be open to pluralism and a diversity of commitments and actions of participating organizations and movements. It welcomes people of all ages, ethnic backgrounds, cultures, generations, and different physical abilities, as long as the participants respect the Charter of Principles. Governments, military organizations, international economics institutions (for example, the IMF, the World Bank, and the WTO), diplomatic representatives, and political parties as such are not allowed to participate. However, government leaders, members of legislatures, and members of political parties are able to participate in a personal capacity if they respect the AWWF Charter.

The 2012 Forum in Marseille was organized as the Forum Alternatif Mondial de l'Eau (FAME) or the AWWF in English. They describe themselves as a concrete alternative to the WWF, which they refer to as a mouthpiece of transnational companies and the World Bank which falsely claims to be the unified voice of global water governance.

The PWF released a Declaration in Istanbul in 2009 to counter the mission of the WWF. The opening paragraph of the Declaration states:

> After Mexico City 2006, which was an important milestone of the continuous work of the global movement for water justice, we have now gathered in Istanbul to mobilize against the 5th World Water Forum. We are here to delegitimize this false, corporate driven World Water Forum and to give voice to the positive agenda of the global water justice movements! (p. 1).

In addition, the PWF noted the devastating impacts of water management policies in Turkey as a powerful example of the expressed concerns. The Declaration demanded that the UN General Assembly organize the next global forum on water. The presence and participation of some UN officials and representatives is noted as "evidence that something has changed" (People's Water Forum Declaration-Istanbul 2009: 1). In addition, the PWF called for making the 2009 WWF the last corporate-controlled water forum. It denounced the WWF's Ministerial Statement "because it does not recognize water as a universal human right and not exclude it from global trade agreements" (People's Water Forum Declaration-Istanbul 2009: 1). The PWF Declaration takes note of the inequity and injustice resulting from promoting the use of water to produce energy from hydroelectric dams. All the issues referenced in the Declaration are in opposition to the WWF's assertions. For instance, the Declaration states:

We reaffirm and strengthen all the principles and commitments expressed in the 2006 Mexico City declaration: we uphold water as the basic element of all life on the planet, as a fundamental and inalienable human right; we insist that solidarity between present and future generations should be guaranteed; we reject all forms of privatization and declare that the management and control of water must be public, social, cooperative, participatory, equitable, and not for profit; we call for the democratic and sustainable management of ecosystems and to preserve the integrity of the water cycle through the protection and proper management of watersheds and environment. (People's Water Forum Declaration-Istanbul 2009: 2)

The PWF (later known as the AWWF) represented the global movement for water justice. This alternative forum, in Istanbul, recognized the interdependence between water and climate change. It committed to "continue building networks and new social alliances, and to involve both local authorities and Parliamentarians who are determined to defend water as a common good and to reaffirm the right to fresh water for all human beings and nature" (People's Water Forum Declaration-Istanbul 2009: 2). It encouraged all public water utilities to get together and establish national associations and regional networks.

Subsequently, in 2012, the AWWF reiterated that the WWC is a mouthpiece for transnational companies and the World Bank, which falsely claim to head the global governance of water. It released a mission statement in Marseille, which advocated pursuing and amplifying the water movement by: (a) creating and promoting an alternative vision of water management which is based on ecological and democratic values; (b) continuing research to find solutions to the worldwide water crisis; and (c) making the water movement structure sustainable. The details of the role of the AWWF were articulated in statements and a charter of action was also released.

The 2012 Charter notes that the AWWF (or FAME in French) is an open meeting place for reflective thinking, the democratic debate of ideas, the formulation of proposals, and the free exchange of experiences. These discussions may lead to effective action and civil society movements which oppose the water resources being managed using profit logic by capitalistic companies and even sometimes by public companies. The AWWF stands for ecological, social, and citizen-based water management, water resource protection, and proper water distribution among different types of users.

The AWWF brings together and interlinks different organizations and movements from civil society from around the world. However, it does not claim to be the voice of all people who fight for water resource protection and the ecological and citizen management of water. It is part of the World Social Forum in the sense that that both forums encourage local and international initiatives to make their voices heard in international organizations. Both forums support movements which inspire change, put transformative actions on the global agenda, and build a better world.

The alternatives proposed by the AWWF stand in opposition to the capitalistic globalization of water and sanitation services which have been implemented by large multinational companies, governments, and international institutions. They serve their own interests, as does the WWF. The AWWF laid out eight key items as representing its views and points or action:

- Water should be a recognized as a common good for all of humanity. Water is vital for all life and is not a commodity.
- National constitutions should officially recognize that all citizens have the right to drinking water and sanitation. This is in accordance with the UN Declaration on 29 July 2010 that access to water is a "human right that is essential for the full enjoyment of life and all human rights."
- Public services should manage water in a participatory manner and protect water resources from agricultural pollution, industrial pollution, pollution from medication, and over-exploitation.
- Water should be democratically distributed among different users in consultation with all parties: the general public, agriculture, industry, and the preservation of biodiversity.
- Necessary infrastructures must be created or improved by government bodies, as they hold public-service authority. These infrastructures must be available everywhere and not only in places where it is profitable; they must distribute and treat water effectively and democratically, while respecting quality norms and protecting the environment.
- Domestic water consumption should be priced using progressive rates to make water use really affordable (and abuse of this should be penalized) and without any profit for the capital invested.

- Alternative techniques in producing and treating water should be promoted. Examples of this are: rainwater recovery, wastewater recycling, lagooning, recovery of humidity in the air, use of morning mist, manual pumping, and solar energy use. Unpatented solutions should be privileged.
- Consequences of global warming should be anticipated: flooding, drought, and freshwater contamination by saltwater. The resilience of ecosystems and soils should be developed.

The AWWF in Marseille 2012 demonstrated that the activities of these Forums are somewhat localized. With the certainty that "another world is possible," as proclaimed in Porte Alegre, the Forum was described as a permanent process of seeking and building alternatives. This process is not reduced to the events supporting it. In an effort to ensure deliberations and a consensus on issues of concern to the AWWF, it noted that no participating organization would be authorized to express positions on behalf of the AWWF if these positions have not been adopted by all participants. Organizations, or groups of organizations, which participated in AWWF meetings were assured the right to deliberate on their own declarations and their own actions, whether single-handedly or in coordination with other participants.

From April 12 to 14, 2015 Korean civil society organizations hosted an Alternative World Water Forum in Daegu in opposition to the corporate-led seventh WWF being held in the same city. One of the key concerns for the Korean Government Employees' Union (KGEU)—the main organizers of the event—was the involvement of the French multinational water corporation, Veolia. KGEU argues that Veolia used the corporate WWF to cement its interests in the region. The KGEU is fighting to keep water public in Korea and represents 140,000 public sector employees, but has been refused legal status by the Korean government. The 2015 Daegu Gyoungbuk declaration on the HRW was released at the Forum. The Declaration covered four main points: the WWF is illegitimate; water and sanitation are human rights; water is part of the commons; and calling for the post-2015 Development Agenda to recognize water and sanitation as human rights and as part of the commons. The Declaration notes that unless the post-2015 Development Agenda is rooted in a human rights framework and a commons-based perspective, it will run the risk of facilitating the commodification of water resources and the privatization of services.

The AWWF is committed to widely circulating such decisions, using the means at its disposal, without directing, hierarchizing, censoring, or restricting them. They are to be taken as deliberations of the particular organization or group of organizations in question. Consider, for instance, the International Indigenous Water Declaration, which is a testament to the Indigenous people's connections to water and expresses the significance of their knowledge and interests to the security of freshwater when water laws and systems are merging into an industry that portrays water as a commodity. In August 2008, a group of Indigenous peoples from across the world convened in northern Australia at the site of the Garma Festival in north-east Arnhem Land to share their experiences on issues and opportunities arising from emerging trends in mainstream water management systems.[12] Noting the long-standing relations between the lives of the Indigenous and water and land, the Declaration's preamble asserts that "water has a right to be recognized as an ecological entity, a being with a spirit and must be treated accordingly."

The Forums also target the state as institutions that enable privatization and trample on the rights of the Indigenous. "Nation-States, in asserting competing sovereignty over the lands and waters, have introduced and enforced unlawful and unjust mechanisms resulting in trespass of the legal entitlements of Indigenous Peoples to the ownership, use, management and benefit of the lands and the waters, without consultation, consent or just compensation where required by law." In addition, the Declaration of the Indigenous calls on states to fully adopt, implement, and adhere to international instruments that recognize their right to land and water.

Similarly, several activists and civil society groups in India signed a declaration as early as 2003—the National People's Water Forum Declaration (in Hindi called the *Jal Suraksha Adhikar, Mukti* Declaration). This Forum, comprising the *Jal Swaraj Abhiyan* and *Rashtriya Jal Biradari*, rejects the privatization agenda articulated in India's New Water Policy and demands the inclusion of "community rights" as the foundation of policy. The Declaration condemns the use of the *panchayat* (the local village governance institution) and NGOs to dismantle public systems and create water users' association and *pani* (meaning water) *panchayats*. Noting that water is a "state" (states of India) subject and that therefore states can legislate on water, the Declaration recognizes the variations in power exerted across national and sub-national state institutions (cf. Subramaniam 2014). The above two examples also point to a wide variety of networks tied to the AWWF. Many of these are nationally and regionally oriented, allowing for ties with locally based struggles.

In anticipation of the 2018 AWWF in Brazil, the announcement for people's participation refers to the planned WWF, which has a budget of millions at its disposal. The AWWF's call reiterates the aims and goals from earlier declarations in relation to their opposition to privatization and the importance of treating water as a common good. In a more specific way, the AWWF emphasizes inequality and gender-based inequities in terms of access to water, particularly the enormous number of hours women and girls spend to collect water. Focusing on the local, the water crisis in Brazil is discussed, as is the negligence of the state in ensuring basic sanitation for people.

There is little evidence of the political influence of the AWWF itself in the sense of directly influencing who should have the power to control rights to water. The lack of funds no doubt limits the ability of the AWWF to disseminate reports as is done by the WWF. In addition, sharing information in a wide variety of languages spoken by activists and civil society groups is a challenge. This is plausibly one reason (the other being resources) why the AWWF typically relies on regional networks (based in the location of the Forum) rather than wider global representation. Unlike the WWF, they do not charge fees for attending sessions. However, the organizations participating in the AWWF have actively supported local struggles to demand rights to water and have opposed private corporations which have used water that local communities have relied on or the privatization of water resources. An example is the role played by the NGO Tarun Bharat Sangh (TBS) in supporting village residents to manage local water resources and facilitating consultative processes among the residents to make decisions. There are also instances of citizens influencing local political institutions to rescind orders for privatizing water resources, such as in Stockton, CA, or accessing groundwater, such as in Plachimada.

GLOBAL INITIATIVES FOR THE CONSCIOUSNESS OF WATER RIGHTS

The global water justice movement is not confined to the AWWF. Anti-privatization groups have organized around the world, including the Africa Water Network, formed in 2007 and now comprising more than 40 African countries. This network has ties to Red VIDA (the International Network for the Defense of the Right to Water) in Bolivia as well as international groups such as Water Justice and the Reclaiming Public Water Network (Gies 2009). These alliances are important for pressuring global financial institutions, private corporations, and state institutions, as well as engaging in legal battles, such as in South Africa and in Plachimada, as discussed in Chapter 3. Other efforts such as the less action-oriented World Water Week and the more vibrant protests against Coca-Cola worldwide have made it possible to raise consciousness about water rights.

In an attempt to continue to recognize the value of water, World Water Week is the annual focal point for the world's water issues.[13] Organized by the Stockholm International Water Institute (SIWI), World Water Week provides a unique forum for the exchange of views, experiences, and practices between the scientific, business, policy, and civic communities. Experts, practitioners, decision-makers, business innovators, and young professionals from a range of sectors and countries come to Stockholm to network, exchange ideas, foster new thinking, and develop solutions to the most pressing water-related challenges of today. A key objective of World Water Week is to track water in the implementation of the 2030 Agenda for Sustainable Development. Each year in World Water Week, decision-makers will have the opportunity to take stock of water's role in the implementation of the water-related Sustainable Development Goals (SDGs) and the Paris Climate Agreement, with the aim of ensuring that water is part of the solutions.

In 2016, over 3200 individuals and around 330 convening organizations from 130 countries participated in World Water Week. At the time of writing, the World Water Week of 2017 has just occurred in Stockholm (August 27–September 1). World leaders, water experts, development professionals, and CEOs met to discuss the world's growing water challenges and how to address them. The theme for the 2017 World Water Week was "Water and Waste: Reduce and Reuse," which resonates with an increasing number of people around the world. A growing global population and more unpredictable weather patterns will increase uncertainty on the availability and quality of water. This has been felt, for example, through a prolonged drought in California, unusually high temperatures and droughts in southern Europe, and a devastating and deadly lack of rain in the Horn of Africa. World Water Week focused special attention on how to mitigate the growing water uncertainty in many parts of the world, discussing how to develop and sustain both technologies and behavior that helps people thrive in an increasingly water-scarce future (see Appendix 4.3 for the various themes covered in the 2017 World Water Week).

Current complex water challenges were addressed by some 3000 participants from nearly 130 countries, representing governments, the private sector, multilateral organizations, civil society, and academia. Speakers at the session included Peter Thomson, President of the UN General Assembly, as well as the Prime Minister of Australia and other influential people. In 2018, World Water Week will address the theme "Water, Ecosystems and Human Development." These educative and informative initiatives occur alongside the vibrant and media-savvy protests against Coca-Cola.

The Campaign to Stop Killer Coke is a worldwide movement based on the efforts of thousands of volunteers.[14] It is described as a movement of "We's and not I's" with a goal of obtaining justice. Ray Rogers is the director of the Campaign. His organization, Corporate Campaign, Inc., created the Campaign to Stop Killer Coke to hold the Coca-Cola Company, its bottlers, and subsidiaries accountable, and to end the gruesome cycle of violence and collaboration with paramilitary thugs, particularly in Colombia. Their campaign spans countries across the world, from South America to Asia to the Middle East.

Protests against the over-extraction of groundwater in India and Sri Lanka by Coca-Cola's subsidiary companies are impacting the parent company. Strong concerns dominated the company's annual general meeting on April 19, 2006 in Delaware, United States. A group of protesters shouted outside the meeting, waving banners with messages such as: "Coca-Cola: Stop De-Hydrating the World" and "Coca-Cola: Destroying Lives, Livelihoods and Communities."

Inside the meeting, nearly 20 shareholders spoke on behalf of campaigns from India and Colombia. A proposal tabled by a shareholder called on the company to "prepare a report on the potential environmental and public health damage of each of its plants, affiliates and proposed ventures extracting water from areas of water scarcity in India," but failed to receive any positive response from the company (see case of Plachimada discussed in Chapter 3). The Coca-Cola Board noted that it understands the need and desire for transparency in all matters including environmental safety and health issues related to the company's operations in India and elsewhere. In doing so, the Board rejected the need for any report as it would lead to a redundant use of Company human and financial resources. *Corporate Accountability International* has actively led a campaign which states that the bottled water industry (involving companies such as Coca-Cola, Pepsi, and Nestlé) is altering people's opinion of bottled water and is creating a demand for a resource that is already available to people for free (Gies 2009).

The campaign against Coca-Cola has spread, particularly on college and university campuses, as well as among trade unionists and religious organizations. The India Resource Centre published a press release the day following the Shareholder meeting stating: "Even as Coca-Cola officials were trying to deal with the scores of protesters at its meeting, the campaign to hold Coca-Cola accountable was producing damning results for the company. Immediately following this, the Union Theological Seminary in Manhattan, New York, a graduate school of theology which trains students to be ministers in the Christian faith, announced that it was banning the sale of Coca-Cola products on its campus."

In India, a new campaign was announced in Gangaikondan, in the southern state of Tamil Nadu, against a Coca-Cola bottling plant under construction. And a massive rally was planned in Plachimada, Kerala on April 22, 2006, where Coca-Cola's bottling plant had remained shut down for over a year because the village council has refused to renew Coca-Cola's license to operate. In November 2006, the Chairman and CEO of the Coca-Cola Company, E. Neville Isdell, spoke about the challenges to Coca-Cola in India at the Nature Conservancy in Atlanta, Georgia. He remarked:

> In India, we have been challenged to demonstrate our commitment to water stewardship. While we are not even close to being one of the largest users of water, we are certainly one of the most visible, and have been subject to criticism that we are depleting groundwater aquifers in the State of Kerala. Let me be very clear: Coca-Cola has a shared interest with the communities where we operate in healthy watersheds—because they sustain life and our business. And the last thing we would ever do is spend millions of dollars to build a plant that would run itself dry. Accordingly, we are working with many partners across India to improve watershed management, and with the Central Ground Water Authority, local governments and communities to expand the use of simple and effective rainwater harvesting technology. To date, we have installed rainwater harvesting systems in 200 locations, including schools and farms, that are helping recharge aquifers when the rains come.

Such conflicts between private entities, the state, and NGOs and local citizens continue in the drive toward the elusive goal of rights to water. Who governs common resources such as water continues to be debated, adopted, and rejected. As global protests intensify and networks such as the WWF and the AWWF fail to have dialogues, the deliberations will remain in separate realms.

There is little evidence from the above discussion to support the claim that there is a shift from "government to governance" that is associated with a transfer of power to other parties within policy-level arenas. Governments are developing networks to engage other governments and other agencies, not to bring nongovernmental stakeholders into the decision-making arena. While greater engagement of nongovernment stakeholders has occurred over time, this is generally facilitated through consultative structures and processes. Governance networks are being used to supplement rather than replace traditional hierarchical government decision-making processes (Jordan et al. 2005).

Water Justice and the State

The state has been at the center of issues of rights to water irrespective of whether we focus on localized struggles or global protests and forums. And in many countries, civil society groups and localized struggles are targeting their governments for support rather than waiting for the UN to enforce international treaties. What have national governments proposed and/or implemented?

The world water movement, in its many configurations, has given birth to national water rights networks (such as in Italy), continental networks (in Africa), and has led to the decline of water privatization in Latin America, Africa, Europe, and notably, in France. France, home to the main multinational water companies, yet has had water come back under public management in Paris.

The water movement has participated in the fight to have access to water recognized as a fundamental human right. Water was declared a human right by the UN on July 28, 2010. In October 2004, the citizens of Uruguay became the first in the world to vote for the right to water. Led by Adriana Marquisio and Maria Selva Ortiz of the National Commission for the Defence of Water and Life, and Alberto Villarreal of Friends of the Earth Uruguay, the groups first had to obtain almost 3000 signatures on a plebiscite (which they delivered to parliament as a "human river") in order to get a referendum placed on the ballot of the national election, calling for a constitutional amendment on the right to water. They won the vote by an majority of almost two-thirds, an extraordinary feat considering the fear-mongering that opponents mounted. The language of the amendment is very important. Not only is water now a fundamental human right in Uruguay, but social considerations must also now take precedence over economic considerations when the government makes water policy. The Constitution clearly notes that the water supply services for human consumption will be provided by the state and not by corporations (Barlow 2008). The Netherlands is the only other country to have passed a law in 2003 restricting the delivery of drinking water by state-owned utilities, but without mentioning the right to water. So only the Uruguayan Constitution guarantees are a model for other countries. In fact, the private company Suez was compelled to leave Uruguay as a result of this amendment.

Several other countries have also passed right to water legislation (see Appendix 4.4). After the end of apartheid in South Africa, Nelson Mandela included water as a human right in the new constitution. However, as it did not include any clause about water services, the World Bank convinced the state to privatize water services, causing much concern among the poor who were unable to pay (Desai 2002). Similarly, other developing countries such as Ecuador, Ethiopia, and Kenya also have references to water as a right in their constitutions, but they also do not include specifics about state services. The Belgian Parliament adopted a resolution in 2005 to recognize water as a human right, as did the French Senate, but neither refers to service/delivery of water.

Article 21 of the Indian Constitution protects citizens' rights to the use of air, water, and earth (the protection of life and personal liberty). It further states that the environmental balance is to be maintained and whenever groundwater is required for domestic and agricultural needs, priority is to be given to these (Venugopal 2006). In 2006, India's Supreme Court ruled that the protection of natural lakes and ponds is akin to honoring the right to life that is a fundamental right to life. In Nepal, activists are demanding the inclusion of the right to water with the right to health.

Ecuador and South Africa are seeing additional demands in terms of gaining rights to water. While celebrating their victory over the privatization of water, the Coalition in Defense of Public Water in Ecuador is seeking a constitutional amendment to recognize the right to water. Similarly, in South Africa, the Coalition Against Water Privatization is challenging the practice of water metering in the Johannesburg High Court, noting that it violates the rights of Soweto's citizens. The President of Bolivia has called for a convention in South America to make water a human right and to reject the market model incorporated in trade agreements. Similar initiatives are evident in other countries such as Colombia and Mexico. The momentum for establishing a right to water is gaining ground across the world.

Conclusion

> We will continue to support each other's struggles as a global water justice movement. We will strengthen our ties of solidarity in our struggles to stop privatisation and promote quality public water and sanitation services at the community level, the national level and the global level.

The above quote is drawn from the 2015 Daegu Gyoungbuk (Korea) declaration on the HRW of the AWWF. It sums up the main focus and mission of the AWWF, which is in direct contrast to the WWF. The 2010 UN Resolution on the HRW has opened up new discussions about what this may imply for people around the world. While some see promise in the UN Resolution, as it can hold states accountable, others note that it may not attend to the needs of the marginalized (such as across class and race and even the Indigenous). Activists call for ongoing conversations about the meaning and implementation of the HRW because "rights" talk are viewed as biased toward individuated rights. Moreover, such an emphasis could leave marginalized communities with sub-standard access to water.

The discussion in this chapter suggests that there is an impasse between the WWF and the AWWF in terms of determining who is involved in the discourse on water rights and what that means. There is a big gap between the mission and goals and the organizations/groups involved in these forums as members and/or participants. As noted above, the right to water is articulated varyingly in these forums. Moreover, these groups differ in their perspectives to address the water crisis, which implies different policy prescriptions and on-the-ground actionable strategies.

The search for documents of the WWC/WWF and the AWWF itself reveal differences in what is articulated about water rights and how. The WWF is a large-scale operation that has a wide variety of members, including private water corporations, global financial institutions, and national governments. The 1997 and 2000 WWF included NGOs and activists as if their mission and that of private water corporations, the state, and global institutions were the same. The PWF/AWWF was only formed in 2003. The massive resources available to the WWF allow it to exert power and dictate the discourse of privatization as the efficient and effective means to address the water crisis. The power of the WWF is also evident in the ability of the WWC to publish reports and organize meetings in an extensive and elaborate way. As noted above, the fees are very high, precluding attendance by activists and community groups, particularly those from the developing world. Access to the WWF is restricted as the proceedings are typically in English (as are the documents). This is in direct contrast to the AWWF meetings, which are small in scale. Even its web presence and information is limited. This is to be expected as the AWWF has few resources.

The broader global water justice movement has enabled the building of alliances and collaborations across locations and these are paying off, particularly as they make it possible to increase the pressure on state institutions. Yet the process of how to accomplish the right to water and what that outcome will look like—practical and realistic mechanisms—needs to developed through dialogue. The battle over water rights and the privatization of water resources is likely to continue.

Notes

1. See www.worldwatercouncil.org (date accessed February 7, 2018).
2. The terms "covenant," "treaty," and "convention" are used interchangeably at the UN.
3. A General Comment is an authoritative interpretation of a human rights treaty or convention. In this case, the interpretation applies to the International Covenant on Economic, Social and Cultural Rights. General Comment Number 15 is therefore an authoritative interpretation that water is a right and an important milestone on the road to a full binding UN convention.
4. www.worldwatercouncil.org/about-us/background (date accessed February 7, 2018).

5. www.worldwatercouncil.org/about-us/vision-mission-strategy (date accessed February 7, 2018).
6. www.worldwatercouncil.org/members/membership-information (date accessed February 7, 2018).
7. www.worldwatercouncil.org/forum/marrakesh-1997 (date accessed February 7, 2018).
8. www.indiaresource.org/news/2006/1032.html (date accessed February 7, 2018).
9. www.freshwateraction.net/content/4th-world-water-forum-mexico-city (date accessed February 7, 2018).
10. www.internationalrivers.org/resources/turkey-deports-international-rivers-staff-after-peaceful-world-water-forum-protest-3806 (date accessed February 7, 2018).
11. www.foeeurope.org/protest-water-forum-140312 (date accessed February 7, 2018).
12. This exchange arose out of relationships between the North Australian Indigenous Land and Sea Management Alliance's Indigenous Water Policy Group and the United Nations University—Institute of Advanced Studies Traditional Knowledge Institute. It was anticipated that access to international experience and perspectives would broaden the frame of reference for participants in such a way as to enhance their capacity to identify and advocate around Indigenous interests in water. The exchange was successful in serving this purpose for all participants from Australia and around the world.
13. Details of World Water Week can be found at: www.worldwaterweek.org.
14. Details of the Campaign can be found at: http://killercoke.org/about.php.

Appendix 4.1

Procedural Rights Codified in International Instruments

	Right to participation	Right to information	Right to effective remedy
Universal Declaration of Human Rights (1948)	Art. 21	Art. 19	Art. 8
American Declaration of the Rights and Duties of Man (1948)	Art. XX	Art. IV (freedom of expression)	Art. XVII
European Convention (1950)[a]	Art. 10	Art. 13	

(continued)

Appendix 4.1 (continued)

	Right to participation	Right to information	Right to effective remedy
CERD (1965)[b]	Art. 5	Art. 5(d)(vii) (freedom of Expression)	Art. 6
ICCPR (1966)[c]	Art. 25	Art. 19	Art. 2(3)
CEDAW (1979)[d]	Art. 7	Arts. 14(2)(b), 16(1)(e)	
African Charter on Human and People's Rights (1981)	Art. 13	Art. 9	Art. 7
World Charter for Nature (1982)	Art. 23	Art. 21(a)	Art. 23
Convention on the Rights of the Child (1990)	Art. 12	Art. 13	
Rio Declaration on Environment and Development (1992)	Pr. 10	Pr. 10	Pr.13
Agenda 21 (1992)	Ch. 23	Ch. 8	Ch. 8.18
Draft Declaration of Principles on Human Rights and the Environment (1994)	Art. 18	Art. 15	Art. 20
IUCN Draft Covenant on Environment and Development (1995)	Art. 12(4)	Art. 12(3)	Art. 12(5)

[a]European Convention for the Protection of Human Rights and Fundamental Freedoms
[b]International Convention on the Elimination of All Forms of Racial Discrimination
[c]International Covenant on Civil and Political Rights
[d]Convention on the Elimination of All Forms of Discrimination against Women
Source Scanlon et al. (2004)

APPENDIX 4.2

Documents of Global Forums (WWF and AWWF)

World Water Council (http://worldwatercouncil.org)
Main webpage, "What's New?" section

1. 2013 Annual Report (www.worldwatercouncil.org/fileadmin/world_water_council/documents/official_documents/WWC_2013AnnualReport.pdf)
2. Access to Water for All, 6th World Water Forum Initiative Report (www.worldwatercouncil.org/fileadmin/world_water_council/documents/official_documents/WWC_AccesstoWaterforAll_Report_June2014.pdf)

3. WWC Position Paper on Water and the Post-2015 Framework (www.worldwatercouncil.org/fileadmin/world_water_council/documents/programs_hydropolitics_sdgs/WWC_Position_Paper_on_water_and_SDGs_Final.pdf)
4. 7th World Water Forum—First Announcement (http://worldwaterforum7.org/File/AttachFileUpload/Community/publications/130924_First_Announcement.pdf)

Forums

1. First World Water Forum "Vision for Water, Life and the Environment"—Marrakesh, 1997
    a. No documents available
2. Second World Water Forum "From Vision to Action"—The Hague, March 2000
    a. World Water Vision Report (presented by the World Commission on Water for the 21st Century) (www.worldwatercouncil.org/library/archives/world-water-vision/vision-report/)
    b. Ministerial Declaration of The Hague on Water Security in the 21st Century (www.worldwatercouncil.org/fileadmin/world_water_council/documents/world_water_forum_2/The_Hague_Declaration.pdf)
3. Third World Water Forum "A Forum with a Difference"—Kyoto, Shiga and Osaka, 2003
    a. Final Report of the 3rd World Water Forum, published by Secretariat of the 3rd World Water Forum 2003 (www.worldwatercouncil.org/fileadmin/world_water_council/documents/world_water_forum_3/3d_World_Water_Forum_FinalReport_BD.pdf)
    b. Analysis of the 3rd World Water Forum, published by the WWC and the Secretariat of the 3rd World Water Forum 2004 (www.worldwatercouncil.org/fileadmin/world_water_council/documents/world_water_forum_3/3d_World_Water_Forum_analysis.pdf)
    c. "Financing Water for All," chaired by Michel Camdessus, written by James Winpenny, published by WWC, Secretariat of the 3rd World Water Forum and Global Water Partnership (www.worldwatercouncil.org/fileadmin/world_water_council/documents_old/Library/Publications_and_reports/CamdessusReport.pdf)

d. "World Water Actions: Making Water Flow for All," written by François Guerquin, Tarek Ahmed, Mi Hua, Tetsuya Ikeda, Vedat Ozbilen, and Marlies Schuttelaar, published by the WWC, Japan Water Resources Association, United Nations Educational, Scientific, and Cultural Organization 2003 (www.worldwatercouncil.org/fileadmin/world_water_council/documents/wwa_overview_en.pdf) (also available in French, Spanish, Chinese, and Japanese)
4. Fourth World Water Forum "Local Action for a Global Challenge"—Mexico, 2006
   a. Summary of the Fourth World Water Forum, published by WWC and the Secretariat of the Fourth World Water Forum (www.worldwatercouncil.org/fileadmin/world_water_council/documents_old/World_Water_Forum/WWF4/synthesis_sept06.pdf) (also available in Spanish)
   b. Final Report of the Fourth World Water Forum, published by WWC and the Secretariat of the Fourth World Water Forum (www.worldwatercouncil.org/fileadmin/world_water_council/documents/world_water_forum_4/Final_Report_4th_Forum.pdf)
   c. Synthesis Process, Phase 1: Session Outcomes
      1. Framework Themes, 1: Water for Growth and Development
         (a) Baseline document (www.worldwatercouncil.org/fileadmin/world_water_council/documents_old/World_Water_Forum/WWF4/Thematic_process/Water_for_Growth_and_Development.pdf)
         (b) Voices of the Forum, Day 1 (www.worldwatercouncil.org/fileadmin/world_water_council/documents_old/World_Water_Forum/WWF4/Voices_of_the_Forum/Voices_of_the_Forum_DAY_1.pdf)
      2. Framework Themes, 2: Implementing Integrated Water Resources Management
         (a) Baseline document (www.worldwatercouncil.org/fileadmin/world_water_council/documents_old/World_Water_Forum/WWF4/Thematic_process/Implementing_Integrated_Water_Resources_Management.pdf)
         (b) Voices of the Forum, Day 2 (www.worldwatercouncil.org/fileadmin/world_water_council/documents_old/World_Water_Forum/WWF4/Voices_of_the_Forum/Voices_of_the_Forum_DAY_2.pdf)

3. Framework Themes, 3: Water Supply and Sanitation for All
   (a) Baseline document (www.worldwatercouncil.org/fileadmin/world_water_council/documents_old/World_Water_Forum/WWF4/Thematic_process/Water_Supply_and_Sanitation_for_All.pdf)
   (b) Voices of the Forum, Day 3 (www.worldwatercouncil.org/fileadmin/world_water_council/documents_old/World_Water_Forum/WWF4/Voices_of_the_Forum/Voices_of_the_Forum_DAY_3.pdf)
4. Framework Themes, 4: Water for Food and the Environment
   (a) Baseline document (www.worldwatercouncil.org/fileadmin/world_water_council/documents_old/World_Water_Forum/WWF4/Thematic_process/Water_Management_for_Food_and_the_Environment.pdf)
   (b) Voices of the Forum, Day 4 (www.worldwatercouncil.org/fileadmin/world_water_council/documents_old/World_Water_Forum/WWF4/Voices_of_the_Forum/Voices_of_the_Forum_DAY_4.pdf)
5. Framework Themes, 5: Risk Management
   (a) Baseline document (www.worldwatercouncil.org/fileadmin/world_water_council/documents_old/World_Water_Forum/WWF4/Thematic_process/Risk_Management.pdf)
   (b) Voices of the Forum, Day 5 (www.worldwatercouncil.org/fileadmin/world_water_council/documents_old/World_Water_Forum/WWF4/Voices_of_the_Forum/Voices_of_the_Forum_DAY_5.pdf)
6. Cross-Cutting Perspectives, 1: New Models for Financing Local Water Initiatives
   (a) Baseline document (www.worldwatercouncil.org/fileadmin/world_water_council/documents_old/World_Water_Forum/WWF4/Thematic_process/New_Models_for_Financing_Local_Water_Initiatives.pdf)
7. Cross-Cutting Perspectives, 2: Institutional Development and Political Processes
   (a) Baseline document (www.worldwatercouncil.org/fileadmin/world_water_council/documents_old/World_Water_Forum/WWF4/Thematic_process/New_Models_for_Financing_Local_Water_Initiatives.pdf)

8. Cross-Cutting Perspective, 3: Capacity-Building and Social Learning
   (a) Baseline document (www.worldwatercouncil.org/fileadmin/world_water_council/documents_old/World_Water_Forum/WWF4/Thematic_process/Capacity_Development_and_Social_Learning.pdf)
9. Cross-Cutting Perspectives, 4: Application of Science, Technology and Knowledge
   (a) Baseline document (www.worldwatercouncil.org/fileadmin/world_water_council/documents_old/World_Water_Forum/WWF4/Thematic_process/Application_of_Science_Technology_and_Knowledge.pdf)
10. Cross-Cutting Perspectives, 5: Targeting, Monitoring and Implementation Assessment
    (a) Baseline document (www.worldwatercouncil.org/fileadmin/world_water_council/documents_old/World_Water_Forum/WWF4/Thematic_process/Targeting_Monitoring_and_Implementation_Assessment.pdf)

5. Fifth World Water Forum "Bridging Divides for Water"—Istanbul, 2009
   a. Water Crossroads (published in 2009 by the WWC) (www.worldwatercouncil.org/fileadmin/world_water_council/documents_old/World_Water_Forum/WWF5/Water_at_a_Crossroad.pdf)
   b. Global Water Framework (published 2009 by the WWC; Fifth World Water Forum Secretariat; Turkish Foreign Ministry) (www.worldwatercouncil.org/fileadmin/world_water_council/documents_old/World_Water_Forum/WWF5/global_water_framework_part_1_final.pdf—part 1)
   (www.worldwatercouncil.org/fileadmin/world_water_council/documents_old/World_Water_Forum/WWF5/global_water_framework_part_2_final.pdf—part 2)
   c. Top 25 Highlights (www.worldwatercouncil.org/fileadmin/world_water_council/documents_old/World_Water_Forum/WWF5/5th_Forum_Highlights.pdf) (also available in French, Spanish, and Russian)
   d. Ministerial Statement (www.worldwatercouncil.org/fileadmin/world_water_council/documents_old/World_Water_Forum/WWF5/Ministerial_Statement_22_3_09.pdf)

e. Istanbul Water Guide (www.worldwatercouncil.org/fileadmin/world_water_council/documents_old/World_Water_Forum/WWF5/IstanbulWaterGuide_Final_05-03-09.pdf)
f. Istanbul Water Consensus (www.worldwatercouncil.org/fileadmin/world_water_council/documents_old/World_Water_Forum/WWF5/Istanbul_Water_Consensus_Eng_Final.pdf)
g. Heads of State Appeal for Action (www.worldwatercouncil.org/fileadmin/world_water_council/documents_old/World_Water_Forum/WWF5/Istanbul_Declaration_Heads_of_State_on_Water.pdf)
h. Forum Programme (www.worldwaterforum5.org/fileadmin/WWF5/Forum_Programme/WWF5_Forum_Programme_Book.pdf)

6. Sixth World Water Forum, "Time for Solutions"—Marseille, 2012
    a. Post Forum Highlights (www.worldwaterforum6.org/fileadmin/user_upload/pdf/publications_elem/Highlights_web_BD_en.pdf) (also available in French)
    b. Global Water Framework (www.worldwaterforum6.org/fileadmin/user_upload/pdf/publications_elem/global_water_framework.pdf)
    c. "Access to Water for All, Supporting Local Communities' Access to Drinking Water and Sanitation," Report on the Sixth World Water Forum Initiative, published by the WWC, 2014 (www.worldwatercouncil.org/fileadmin/world_water_council/documents/official_documents/WWC_AccesstoWaterforAll_Report_June2014.pdf)
    d. Pre Forum Publications
        1. Pocket Programme (www.worldwaterforum6.org/fileadmin/user_upload/pdf/programme_elem/Programme-6thWorldWaterForum-6emeForumMondialdelEau.pdf) (also available in Russian)
        2. World Water Priorities (www.worldwaterforum6.org/fileadmin/user_upload/pdf/programme_elem/Programme-6thWorldWaterForum-6emeForumMondialdelEau.pdf)
        3. Contribute to "Time for Solutions" (www.worldwaterforum6.org/fileadmin/user_upload/pdf/Thematic_targets_EN.pdf)
        4. Solutions Kit (www.worldwaterforum6.org/fileadmin/user_upload/pdf/Solutions_for_Water.pdf)

5. Second Announcement (www.worldwaterforum6.org/fileadmin/user_upload/pdf/second_announcement.pdf)
6. First Announcement (www.worldwaterforum6.org/wp-content/uploads/2011/04/1st-Announcement.pdf)
7. Become a Volunteer for the Forum (www.worldwaterforum6.org/fileadmin/user_upload/pdf/ForumMondialdelEau-Depliant-benevoles-FR.pdf)
8. O'Pen Pack (www.worldwaterforum6.org/fileadmin/user_upload/pdf/Pack_O_Pen.pdf)
9. Leaflets Processes (www.worldwaterforum6.org/wp-content/uploads/2011/04/Leaflet-process-English.pdf) (also in available in French, Chinese, Turkish, Arabic, Spanish, and Russian)

7. Seventh World Water Forum "Water for Our Future"—Daegu and Gyeonbuk, 2015
   a. First Announcement (www.worldwatercouncil.org/fileadmin/world_water_council/documents/official_documents/2014-06_Forum7-First_Announcement.pdf)
   b. Second Announcement (www.worldwatercouncil.org/fileadmin/world_water_council/documents/world_water_forum_7/Forum7-2nd%20Announcement.pdf)
   c. Water for our Future, Presentation Brochure (www.worldwatercouncil.org/fileadmin/world_water_council/documents/world_water_forum_7/Forum7_water_for_our_future.pdf)
   d. Thematic Framework (www.worldwatercouncil.org/fileadmin/world_water_council/documents/world_water_forum_7/The_Thematic_Framework_of_the_7th_Forum_.pdf)

Water Prizes

1. King Hassan II Great World Water Prize
   a. Brochure (www.worldwatercouncil.org/fileadmin/world_water_council/documents/King_Hassan_II_Great_World_Water_Prize/HassanII-brochure-formulaire-EN.pdf)
2. Kyoto World Water Grand Prize
   a. Guidelines (www.worldwatercouncil.org/fileadmin/world_water_council/documents/Kyoto_world_water_grand_prize/20140820%20Guidelines%20for%20Kyoto%20World%20Water%20Grand%20Prize_checked_re.pdf)

b. Application Form (www.worldwatercouncil.org/fileadmin/world_water_council/documents/Kyoto_world_water_grand_prize/4th-Kyoto-prize_application.pdf)

Programs—Water and Energy

1. Water for Energy, World Energy Council, 2010 (www.worldwatercouncil.org/fileadmin/world_water_council/documents/programs_hydropolitics_sdgs/water_energy_1.pdf)
2. The Status of the Water-Food-Energy Nexus in Asia and the Pacific, United Nations, 2013 (www.worldwatercouncil.org/fileadmin/world_water_council/documents/programs_hydropolitics_sdgs/Water-Food-Nexus%20Report.pdf)
3. The Post-2015 Water—Thematic Consultation, Resources Management Framing Paper, 2013 (www.worldwatercouncil.org/fileadmin/world_water_council/documents/programs_hydropolitics_sdgs/WRM-1-Energy-Framing-Paper.pdf)
4. Water, Energy and Climate Change—A Contribution from the Business Community, World Business Council for Sustainable Development, 2009 (www.worldwatercouncil.org/fileadmin/world_water_council/documents/programs_hydropolitics_sdgs/WRM-1-Energy-Framing-Paper.pdf)
5. Sixth World Water Forum Synthesis Report on the Priority for Action 2.3—Harmonize Energy and Water, Strategic Direction: Contribute to Economic Development, 2012 (www.worldwatercouncil.org/fileadmin/world_water_council/documents/programs_hydropolitics_sdgs/2-3%20%282%29.pdf)

Members

1. Application Form (www.worldwatercouncil.org/fileadmin/world_water_council/documents/wwc-membership/WWC-2012MembershipApplicationForm.pdf)
2. Subsidy Application Form (www.worldwatercouncil.org/fileadmin/world_water_council/documents/wwc-membership/WWC_Subsidy_Application_Form.pdf)
3. List of Members. April 2014 (www.worldwatercouncil.org/fileadmin/world_water_council/documents/wwc-membership/List_of_Members_per_college_April_2014.pdf)

4. Membership Brochure (www.worldwatercouncil.org/fileadmin/world_water_council/documents/wwc-membership/Membership-leaflet_2013_web.pdf)
5. Membership Guidelines (www.worldwatercouncil.org/fileadmin/world_water_council/documents/wwc-membership/WWC_Membership_Guidelines.pdf)
6. Constitution & By-Laws (www.worldwatercouncil.org/fileadmin/world_water_council/documents/Constitution_ByLaws_2013_2015.pdf) (also available in French, Spanish, and Russian).

Activity Reports

1. "Water, the Key for Global Development" Annual Report 2013, published by the WWC, 2014 (www.worldwatercouncil.org/fileadmin/world_water_council/documents/official_documents/WWC_2013AnnualReport.pdf)
2. "A Pact for Water Security" Strategy 2013–2015, published by the WWC, 2013 (www.worldwatercouncil.org/fileadmin/world_water_council/documents/official_documents/2013-05_Strategy_2013-2015.pdf)
3. "Hydro-diplomacy in Motion" Activity Report 2010–2012, published by the WWC, 2012 (www.worldwatercouncil.org/fileadmin/world_water_council/documents/official_documents/WWC_triennal_report_low_res.pdf)
4. "Water, a Global Priority" Annual Report 2010, published by the WWC, 2011 (www.worldwatercouncil.org/fileadmin/world_water_council/documents_old/Library/Publications_and_reports/Activity_reports/Rapport_Annuel_WWC_2010-_GB_WEB.pdf)
5. "Politics Get into Water" Triennial Report 2006–2009, published by the WWC, 2009 (www.worldwatercouncil.org/fileadmin/world_water_council/documents_old/Library/Publications_and_reports/Activity_reports/TriennalReport_2006-2009.pdf)
6. World Water Council Biennial Report 2004–2005, published by the WWC, 2004–2005 (www.worldwatercouncil.org/fileadmin/world_water_council/documents_old/About_us/official_documents/Biennial_report_2004-2005_ENG.pdf) (also available in French)
7. World Water Council Triennial Report 2000–2003, published by the WWC, 2003 (www.worldwatercouncil.org/fileadmin/world_water_council/documents_old/Library/Publications_and_reports/Activity_reports/triennial_2000-2003.pdf) (also available in French)

8. WWC Brochure (www.worldwatercouncil.org/fileadmin/world_water_council/images/publications/WWC_Plaquette_A4_web.pdf)

Thematic Publications

1. "Cooperating Beyond Borders: Successful Examples of Transboundary Water Management," Seminar Proceedings, published by the WWC and the Cercle français de l'eau with the support of the Rhine-Meuse Water Agency and the collaboration of the European Parliament, 2014 (www.worldwatercouncil.org/fileadmin/world_water_council/documents/publications/2014_02_25_Cooperating_beyond_borders_seminar_proceedings.pdf)
2. "The Right to Safe Water and Sanitation—A Priority," flyer, published by the WWC, 2012 (www.worldwatercouncil.org/fileadmin/world_water_council/documents_old/Library/Publications_and_reports/3.Flyer_Right-to-water.pdf) (also available in French)
3. "Water and Green Growth," joint publication by the Government of the Republic of Korea and the WWC, 2012 (www.worldwatercouncil.org/fileadmin/world_water_council/documents_old/Library/Publications_and_reports/2.Green_Growth_ExecutiveSummary.pdf—executive summary) (also available in Spanish) (www.worldwatercouncil.org/fileadmin/world_water_council/documents_old/Library/Publications_and_reports/2.Green_Growth_Report_Edition1.pdf—full report)
4. "Water for Growth and Development in Africa—A Framework for an Effective Mosaic of Investments," published by the WWC, with the support of the French Ministry of Foreign Affairs, 2011 (www.worldwatercouncil.org/fileadmin/world_water_council/documents_old/Library/Publications_and_reports/Africa_Report.pdf) (also available in French)
5. The Water Caucus—Summary of Workshops during the 2009 General Assembly
   a. Climate Challenges (www.worldwatercouncil.org/fileadmin/world_water_council/documents_old/Library/Climate Challenges.pdf)

b. Efficient Footprints (www.worldwatercouncil.org/fileadmin/world_water_council/documents_old/Library/Efficient Footprints.pdf)
c. Urban Urgency (www.worldwatercouncil.org/fileadmin/world_water_council/documents_old/Library/UrbanUrgency.pdf)
d. Transboundary Futures (www.worldwatercouncil.org/fileadmin/world_water_council/documents_old/Library/TransboundaryFutures.pdf)
6. "'Don't Stick Your Head in the Sand': Towards a Framework for Climate-Proofing," published by the WWC, 2009 (www.worldwatercouncil.org/fileadmin/world_water_council/documents_old/Library/Publications_and_reports/Towards_a_Framework_for_Climate-Proofing.pdf)
7. "Perspectives on Water and Climate Change Adaptation," Introduction, summaries and key messages, published by the WWC, 2009 (www.worldwatercouncil.org/fileadmin/world_water_council/documents_old/Library/Publications_and_reports/0._Introduction__Summaries_and_Key_Messages.pdf)
8. "Perspectives on Water and Climate Change Adaptation," Publication Series, published by the WWC; CPWC; IUCN; IWA, 2009
   a. The Changing Himalayas (www.worldwatercouncil.org/fileadmin/world_water_council/documents_old/Library/Publications_and_reports/Climate_Change/PersPap_01._The_Changing_Himalayas.pdf)
   b. Environment as Infrastructure (www.worldwatercouncil.org/fileadmin/world_water_council/documents_old/Library/Publications_and_reports/Climate_Change/PersPap_02._Environment_as_Infrastructure.pdf)
   c. Small Island Countries (www.worldwatercouncil.org/fileadmin/world_water_council/documents_old/Library/Publications_and_reports/Climate_Change/PersPap_03._Small_Island_Countries.pdf)
   d. Planning Better Water Resources Management (www.worldwatercouncil.org/fileadmin/world_water_council/documents_old/Library/Publications_and_reports/Climate_Change/PersPap_04._Planning_Better_WRM.pdf)

e. Producing Enough Food (www.worldwatercouncil.org/fileadmin/world_water_council/documents_old/Library/Publications_and_reports/Climate_Change/PersPap_05._Producing_Enough_Food.pdf)
f. Transboundary Water Management (www.worldwatercouncil.org/fileadmin/world_water_council/documents_old/Library/Publications_and_reports/Climate_Change/PersPap_06._Transboundary_Water_Management.pdf)
g. Local Government (www.worldwatercouncil.org/fileadmin/world_water_council/documents_old/Library/Publications_and_reports/Climate_Change/PersPap_07._Local_Government.pdf)
h. Business (www.worldwatercouncil.org/fileadmin/world_water_council/documents_old/Library/Publications_and_reports/Climate_Change/PersPap_08._Business.pdf)
i. Arid and Semi-Arid Regions (www.worldwatercouncil.org/fileadmin/world_water_council/documents_old/Library/Publications_and_reports/Climate_Change/PersPap_09._Arid_and_Semi-Arid_Regions.pdf)
j. Water Industry (www.worldwatercouncil.org/fileadmin/world_water_council/documents_old/Library/Publications_and_reports/Climate_Change/PersPap_10._Water_Industry.pdf)
k. Financial Issues (www.worldwatercouncil.org/fileadmin/world_water_council/documents_old/Library/Publications_and_reports/Climate_Change/PersPap_11._Financial_Issues.pdf)
l. Energy (www.worldwatercouncil.org/fileadmin/world_water_council/documents_old/Library/Publications_and_reports/Climate_Change/PersPap_12._Energy.pdf)
m. Deltas (www.worldwatercouncil.org/fileadmin/world_water_council/documents_old/Library/Publications_and_reports/Climate_Change/PersPap_13._Deltas.pdf)
n. WASH Services Delivery (www.worldwatercouncil.org/fileadmin/world_water_council/documents_old/Library/Publications_and_reports/Climate_Change/PersPap_14._WASH_Services_Delivery.pdf)

o. Water Resources and Services (www.worldwatercouncil.org/fileadmin/world_water_council/documents_old/Library/Publications_and_reports/Climate_Change/PersPap_15._Water_Resources_and_Services.pdf)

p. IWRM and Sea (www.worldwatercouncil.org/fileadmin/world_water_council/documents_old/Library/Publications_and_reports/Climate_Change/PersPap_16._IWRM_and_SEA.pdf)

9. "Water and Disaster," water and disaster senior expert discussion groups, UN Secretary-General Advisory Board on Water and Sanitation (UNSGAB), published by the WWC, 2009 (www.worldwatercouncil.org/fileadmin/world_water_council/documents_old/Library/Publications_and_reports/Publication_Water_and_Disaster_lowres.pdf)

10. "Creditor Reporting System on Aid Activities 2008" in Support of Water Supply and Sanitation 2001–2006, published by the OECD/DAC Secretariat and the WWC, 2008 (retrieved 01/25/2015, 7:09 p.m.)

11. "Task Force on Financing Water for All—Report 1. Enhancing Access to Finance for Local Governments—Financing Water for Agriculture," chaired by Angel Gurria, written by Paul Van Hofwegen, published by the WWC, 2006 (www.worldwatercouncil.org/fileadmin/world_water_council/documents/publications/Financing_FinalText_Cover.pdf)

12. "The Right to Water—From Concept to Implementation," written by Céline Dubreuil, published by the WWC (www.worldwatercouncil.org/fileadmin/world_water_council/documents_old/Library/Publications_and_reports/RightToWater_FinalText_Cover.pdf) (also available in French and Spanish)

13. "Costing MDG Target 10 on Water Supply and Sanitation: Comparative Analysis, Obstacles and Recommendations," written by Jérémie Toubkiss, published by the WWC, 2006 (www.worldwatercouncil.org/fileadmin/world_water_council/documents_old/Library/Publications_and_reports/FullTextCover_MDG.pdf) (also available in French)

14. "Official Development Assistance for Water from 1990 to 2004—Figures and Trends," written by Florence Clermont, published by the WWC, 2006 (www.worldwatercouncil.org/fileadmin/world_water_council/documents_old/Library/Publications_and_reports/FullText_Cover_ODA.pdf)

15. "Proceedings of the Workshop on Water and Politics Understanding the Role of Politics in Water Management" published by WWC 2004 (retrieved 01/25/2015, 7:19 p.m.)
16. "E-Conference Synthesis: Virtual Water Trade—Conscious Choices," published by the WWC, 2004 (www.worldwatercouncil.org/fileadmin/world_water_council/documents_old/Library/Publications_and_reports/virtual_water_final_synthesis.pdf)

People's Water Forum

1. First People's Water Forum—Florence, March 21–22, 2003
   a. Website for First People's World Water Forum (Italian only) http://contrattoacqua.it
      Info here: http://unterm.un.org/DGAACS/unterm.nsf/8fa942046ff7601c85256983007ca4d8/028e187554462b-0b85256e740069683c?OpenDocument)
   b. "Peoples' Water Forum Miles from World Water Forum," March 4, 2003 (www.ictsd.org/bridges-news/biores/news/peoples-water-forum-miles-from-world-water-forum)
   c. "Peoples World Water Forum—March 21 and 22, 2003, Florence, Italy" n.d. (http://ciel.org/Trade_Sustainable_Dev/WaterForum_24Mar03.html)
   d. "National People's Water Forum Declaration," March 15–16, 2003 (https://www.citizen.org/documents/NationalPeople's WaterForumdeclaration.pdf)
   e. "The People's World Water Forum—Florence, Italy" n.d. (www.citizenarchive.org/cmep/Water/conferences/articles.cfm?ID=9280)
2. Second People's Water Forum—Istanbul, 2009
   a. "Peoples Water Forum Declaration," March 19, 2009 (www.tni.org/article/peoples-water-forum-declaration)
      Pdf also available (http://wilpf.org/files/PeoplesWaterForum Declaration_03-09.pdf)
   b. "Water Rights Activists Blast Istanbul World Water Forum as 'Corporate Trade Show to Promote Privatization,'" March 23, 2009 (www.democracynow.org/2009/3/23/water_rights_activists_blast_istanbul_world)

c. "Water as Commodity or Commons? Issues from the 2009 World Water Forum," May 3, 2009 (www.projectcensored.org/water-as-commodity-or-commons-issues-from-the-2009-world-water-forum)
d. "Opening Message at the People's Water Forum, Istanbul, Turkey" n.d. (http://focusweb.org/node/1477)
e. "A Water Group Protests the World Water Forum: Why We're Part of the People's Water Forum Instead," March 18, 2009 (http://www.foodandwaterwatch.org/blogs/a-water-group-protests-the-world-water-forum-why-were-part-of-the-peoples-water-forum-instead)
f. "World Water Forum, Istanbul March 2009" n.d. (http://worldwaterforum.blogspot.com)
g. "Challenging the World Water Forum to Protect Water from Corporate Control," March 12, 2012 (https://www.stopcorporateabuse.org/press-statement/challenging-world-water-forum-protect-water-corporate-control)

Alternative World Water Forum
Official website: www.fame2012.org/en
"First People's World Water Forum," March 21–22, 2003 (http://water-l.iisd.org/events/first-peoples-world-water-forum)
Announcement for the first Alternative World Water Forum, held in Florence, on March 21–22, 2003.
"People, Planet and Water," March 30, 2012 (http://rio20.net/en/propuestas/people-planet-and-water)
Declaration of principles and proposals by the Alternative World Water Forum, drafted at a gathering on March 9–10, 2012, in Marseille during the WWC's Sixth World Water Forum.
"Alternative World Water Forum" n.d. (http://rio20.net/en/events/alternative-world-water-forum)
Announcement and description of the Alternative World Water Forum to be held March 9–10, 2012, in Marseille during the WWC's Sixth World Water Forum.
"The Resounding Message from People's World Water Forum: Reclaim Public Water!" n.d. (www.world-psi.org/en/resounding-message-peoples-world-water-forum-reclaim-public-water)

Article about the Sixth World Water Forum and the Alternative World Water Forum.
Indigenous Peoples
"Indigenous Peoples' Kyoto Water Declaration," March 2003 (www.waterculture.org/uploads/IPKyotoWaterDeclarationFINAL.pdf)
Third World Water Forum, Kyoto, Japan
"Indigenous Peoples' Contributions," March 16–23, 2003 (http://portal.unesco.org/science/en/ev.php-URL_ID=3855&URL_DO=DO_TOPIC&URL_SECTION=201.html)
"International Indigenous Water Declaration," August 2008 (www.unutki.org/downloads/File/Events/International_Indigenous_Water_Declaration.pdf)

APPENDIX 4.3

World Water Week 2017

Water is key to a range of issues that will shape the world in the decades to come. These were discussed in-depth during World Water Week.

Water and climate: climate change is to a large extent water change. Water disasters account for more than 90% of the natural disasters in the world, and climate-driven water hazards, water scarcity, and variability pose significant risks to all economic activity, such as food and energy production, manufacturing and infrastructure development, as well as political stability. This is also true for high-income countries. Resilience to climate change requires adaptive water management and robust water infrastructure.

Sustainable Development Goals (SDGs): nearly all the SDGs will require water to be achieved, and implementation will need to be integrated and coordinated. Water can help to facilitate this. For example, energy and food security, as well as economic growth and urbanization (SDGs 2, 7, 8, and 11), are directly dependent on the availability of freshwater resources.

Water as connector between the SDGs and the Paris Agreement: in the 2030 Agenda for Sustainable Development, the water and sanitation SDG (Goal 6) links across all the other 16 goals with a large number of water-related targets in the overall Agenda, making water a key underlying factor and entry point for the successful implementation of the 2030 Agenda. For the Paris Agreement, most of the countries who submitted their Nationally Determined Contributions (NDCs) prioritized water in their adaptation chapters over agriculture and health. This positions water as a priority for national policy, program implementation, and funding.

Drinking water and sanitation: the global water and sanitation crisis is mainly rooted in poverty, power, and inequality, not in physical water scarcity. It is, first and foremost, a crisis of governance. Poor resources management, corruption, lack of appropriate institutions, bureaucratic inertia, and insufficient capacity lie in many places behind the lack of sustainability of services, which also undermine the arrival of new investments. Better water governance is needed for enhancing sustainability of services and attracting more investment into the sector.

Water security: to manage the global rise in demand for water and to increase water productivity, incentives for using water more effectively are necessary. Water needs to be given its true value for production purposes in the energy, industrial, and agricultural sectors.

Water and food/nutrition: although prevalence is declining, an estimated 800 million people are still undernourished. A worrying opposite trend is a rapid growth of over-eating: well over two billion are now overweight, obese, or are negatively affected by diets that are less healthy. This kind of New Normal and maldevelopment is a global phenomenon with the most rapid increase among young people, and also notable among the poor.

Innovative financing and green bonds: a great deal of (sustainable and climate-smart) finance will be needed both for supplying water and treating waste water, but these investments to increase resilience to climate change will be much cheaper than the emergency responses which a future changed climate will require in terms of food security and human health. An investment in climate-proof infrastructure today will be offset by a future reduced need for emergency response measures to counter floods and droughts.

Water cooperation: development needs cooperation. Cooperation over transboundary waters would spur regional development, improve resilience to climate change, and decrease the risk of geopolitical hostility.

The political aspects of transboundary cooperation cannot be neglected if real progress is to be made.

Water integrity: corruption is one of the most serious challenges to sustainable management of water resources management and the provision of water services. It reduces economic growth, discourages investment, increases the services delivery costs, increases health risks, and robs poor people of their livelihoods and access to water.

Pricing of water and valuing water: water needs to be better valued. Some parts of this value can easily be reflected in a price, while others cannot. Therefore, water pricing needs to be complemented with other types of policy instruments (such as laws, public awareness-raising, or standards). In particular, we need to make sure that basic water services are also affordable for the poorest people, as per the HRW and sanitation, and that water continues to keep ecosystems healthy.

Water and migration: increasingly, researchers and policy makers are seeking to explain migration and refugee flows in terms of water scarcity—often perpetuated by climate change. The interlinks between water challenges and climate change manifested in the form of, for example, increased variability and uncertainty are not the main causes of large-scale population migration; rather, they should be considered as push factor multipliers. Social, economic, and political factors will also affect the vulnerability or resilience of communities.

Water and faith: water has profound symbolic meaning in many religious and local traditions, and water scarcity is particularly acute in many parts of the world in which faith is a central aspect of individual and community identity. Water governance and development are not only about policies and investments, but are rooted in behavior change and cultural values. In that respect, the role of faith-based organizations becomes crucial, given their presence and influence in local communities.

Pharmaceuticals and water: active pharmaceutical ingredients are micro-pollutants and are of growing concern around the world. Manufactured to be stable enough to reach and interact with the relevant organ, many pharmaceuticals are not easily biodegradable and remain in the environment for considerable periods of time.

APPENDIX 4.4
Countries with Legal Provisions on Water

Country	Legal provision
Bulgaria	Article 55 of the Constitution: citizens have the right to a healthy and favorable environment.
Burkina Faso	Article 31 of the Constitution: the right to a healthy environment shall be recognized.
Cambodia	Article 59 of the Constitution (1993): the State shall protect the environment and the balance of abundant natural resources, and shall establish a precise plan of management of land, water, air, wind, geology, ecological system, mines, energy, petrol and gas, rocks and sand, gems, forests and forest products, wildlife, fish, and aquatic resources.
Cape Verde	Article 70 of the Constitution: everyone shall have the right to a healthy, ecologically balanced environment.
Chile	Article 19(8) of the Constitution (1980): the Constitution guarantees to all persons the right to live in an environment free from pollution.
Colombia	Article 79 of the Constitution: every individual has the right to enjoy a healthy environment.
Congo	Article 46 of the Constitution: each citizen shall have the right to a healthy, satisfactory, and enduring environment.
Costa Rica	Articles 46 and 50 of the Constitution recognizes a right of the people to a healthy environment.
Ecuador	Article 23 of the Constitution recognizes a right of the people to a healthy environment.
Eritrea	Article 10 of the Constitution (1996): the stat e shall work to bring about a balanced and sustainable development throughout the country, and shall use all available means to ensure all citizens to improve their livelihood in a sustainable manner through their development. The state shall have the responsibility to regulate all land, water, and natural resources, and to ensure their management in a balanced and sustainable manner, and in the interests of the present and future generations, and to create the right conditions for securing the participation of the people to safeguard the environment.
Ethiopia	Article 90(1) of the Constitution (1998): every Ethiopian is entitled, within the limits of the country's resources, to clean water. Article 92: the state has a duty to control and utilize land and natural resources for the common good of the nation's people and for their development. The state has an obligation to strive to ensure a clean and healthy environment and other environmental rights.

(continued)

Appendix 4.4 (continued)

Country	Legal provision
France	Paragraph 11 of the Preamble to the French Constitution of 1946: "[The nation] guarantees to all ... health protection, material security" The French Water Act of 1992 says that the use of water belongs to everyone.
Gambia	Article 216(4) of the Constitution (1996): the State shall endeavor to facilitate equal access to clean and safe water.
Greece	Article 24 of the Constitution recognizes a right of the people to a healthy environment.
Guatemala	Article 127 of the Constitution (1985): water regime. All waters belong to the public domain, and are inalienable and imperceptible. The use and enjoyment are granted in the ways established by law and in accordance to social interest. A specific law will regulate this matter. Article 128: use and enjoyment of water, lakes, and rivers. The use and enjoyment of lake and river waters, for agricultural, farming, tourist or purposes of any other nature, that contribute to the development of domestic economy are at the service of the community and not of any particular person, but users shall have the obligation to reforest the shores and channels, as well as to provide access routes.
Honduras	Article 145 of the Constitution: recognizes a right of the people to a healthy environment.
Hungary	Chapters 1 and 18 of the Constitution (1990): the Republic of Hungary recognizes everyone's right to a healthy environment.
India	Article 48A of the Constitution: the stat e shall endeavor to protect and improve the environment and to safeguard the forests and wildlife of the country. Article 51A: it shall be the duty of every citizen of India to protect and improve the natural environment, including forests, lakes, rivers, and wildlife, and to have compassion for living creatures.
Japan	Article 25 of the Constitution: citizens "have the right to maintain the minimum standards of wholesome and cultured living".
Kazakhstan	Article 26 of the Constitution: a citizen shall have the right to a favorable environment.
Korea	Chapter II, Article 35 of the Constitution (1987): all citizens shall have the right to a healthy and pleasant environment.
Kyrgyzstan	Article 35 of the Constitution: citizens of the Kyrgyz Republic shall have the right to a healthy, safe environment.
Laos	Article 17 of the Constitution (1991): all organizations and citizens must protect the environment and natural resources: land, underground, forests, fauna, water sources, and the atmosphere.

(continued)

Appendix 4.4 (continued)

Country	Legal provision
Macedonia (Former Yugoslav Republic)	Article 43 of the Constitution: everyone has the right to a healthy environment to live in.
Mali	Article 15 of the Constitution: every person shall have the right to a healthy environment.
Mexico	Article 27 of the Constitution (amended 1999): ownership of the lands and waters within the boundaries of the national territory is vested originally in the nation, which has had, and has, the right to transmit title thereof to private persons, thereby constituting private property. Centers of population which at present either have no lands or water or which do not possess them in sufficient quantities for the needs of their inhabitants shall be entitled to grants thereof, which shall be taken from adjacent properties, the rights of small landed holdings in operation being respected at all times.
Moldova	Article 37 of the Constitution: every human being has the right to live in an environment that is ecologically safe for life and health.
Namibia	Preamble to the Sixth Draft Water Resources Management Bill (2001): the government's overall responsibility for an authority over the nation's water resources and their use, including the equitable allocation of water to ensure the right of all citizens to sufficient safe water for a healthy and productive life and the redistribution of water.
The Netherlands	Article 21 of the Constitution: it shall be the concern of the authorities to keep the country habitable and to protect and improve the environment.
Nicaragua	Chapter III, Article 60 of the Constitution (1987): Nicaraguans have the right to live in a healthy environment. Article 36 of Law No. 28 (1987): communal property is the lands, waters, and forests that have traditionally belonged to the communities of the Atlantic Coast, and they are subject to the following provisions: 1. Communal lands are inalienable; they cannot be donated, sold, encumbered nor mortgaged, and they are inextinguishable.
Panama	Article 114 of the Constitution: it is a fundamental duty of the state to guarantee that the population live in a healthy environment, free of pollution, where air, water, and food satisfy the development requirements for an adequate development of human life.
Paraguay	Article 7 of the Constitution: everyone has the right to live in a healthy, ecologically balanced environment.

(continued)

Appendix 4.4 (continued)

Country	Legal provision
Peru	Article 123 of the Constitution: everyone has the right to live in a healthy environment, ecologically balanced and adequate for the development of life and the preservation of the countryside and nature.
The Philippines	Articles II and 15 of the Constitution: the state shall protect and advance the right of the people to a balanced and healthful ecology in accordance with the rhythm and harmony of nature.
Portugal	Article 66(1) of the Constitution (1982): everyone shall have the right to a healthy and ecologically balanced human environment and the duty to defend it.
Slovakia	Article 44 of the Constitution: every person has the right to a favorable environment.
Slovenia	Article 71 of the Constitution: everyone shall have the right to a healthy living environment in accordance with the law.
South Africa	§24(a) of the Constitution (1996): everyone has the right to an environment that is not harmful to their health or well-being. The South African Bill of Rights (1996) enshrines a right of access to water in Section 27: 1. Everyone has the right to have access to (a) health care services, including reproductive health care; (b) sufficient food and water; and (c) social security, including, if they are unable to support themselves and their dependents, appropriate social assistance. 2. The state must take reasonable legislative and other measures, within its available resources, to achieve the progressive realization of each of these rights.
Spain	Article 45 of the Constitution: every person shall have the right to enjoy an environment suitable for the development of the person.
Switzerland	Article 24bis of the Constitution (1991): 1. to ensure the economical use and protection of water and the prevention of damage by water, the Confederation, having regard to the total water economy, shall by legislation establish principles in the general interest concerning: (a) the conservation and exploitation of water, especially for the supply of drinking water and the enrichment of underground water; (b) the use of water for energy production and for cooling purposes; (c) the regulation of water levels and of the flow of surface and underground water, the diversion of water outside its natural course, irrigation and drainage and other intervention in the water cycle.
Togo	Article 41 of the Constitution: every person shall have the right to a clean environment.
Turkey	Article 56 of the Constitution (1982): everyone has the right to live in a healthy, balanced environment.

(continued)

Appendix 4.4 (continued)

Country	Legal provision
Uganda	Preamble to the Constitution (1995): the State shall protect important natural resources, including land, water, wetlands, minerals, oil, fauna, and flora on behalf of the people of Uganda. Article 14: the state shall endeavor to fulfill the fundamental rights of all Ugandans to social justice and economic development, and shall, in particular, ensure that all Ugandans enjoy rights and opportunities and access to education, health services, clean and safe water, decent shelter, adequate clothing, food security, and pension and retirement benefits.
United States	Eight state constitutions recognize the right to a healthy environment. Besides, the Constitutions of Illinois, Pennsylvania, Massachusetts, and Texas, all recognize the right of people to pure water.
Venezuela	Article 127 of the Constitution: recognizes the right of people to a healthy environment.
Zambia	Article 112 of the Constitution (1996): the stat e shall endeavor to provide clean and safe water.

Source Scanlon et al. (2004)

REFERENCES

Bakker, K. 2007. "The Commons Versus the Commodity: After Globalization, Anti-privatization and the Human Right to Water in the Global South." *Antipode* 39(3): 430–455.

Barlow, Maude. 2008. "Making Water a Human Right." In *Water Consciousness: How We All Have to Change to Protect Our Most Critical Resources*, edited by Tara Lohan, pp. 177–185. San Francisco, CA: AlterNet Books.

Barlow, Maude, and Tony Clarke. 2002. *Blue Gold: The Fight to Stop the Corporate Theft of the World's Water.* New York: New Press.

Bond, P., and J. Dugard. 2008. "Water, Human Rights and Social Conflict: South African Experiences." *Law, Social Justice & Global Development* 1: 1–21.

Clarke, Tony. 2008. "On Developing 'Water Consciousness': Eight Movement Building Principles." In *Water Consciousness: How We All Have to Change to Protect Our Most Critical Resources*, edited by Tara Lohan, pp. 161–167. San Francisco, CA: AlterNet Books and Watershed Media.

Desai, Ashwin. 2002. *We Are the Poors: Community Struggles in Post-apartheid South Africa.* New York, NY: Monthly Review Press.

Fifth World Water Forum Highlights from Istanbul. 2009. Accessed January 25, 2015. http://www.worldwatercouncil.org/fileadmin/world_water_council/documents_old/World_Water_Forum/WWF5/5th_Forum_Highlights.pdf.

Food & Water Watch. 2007. *The Price of Privatization.* Stockton, CA: Report

Fourth WWF Final Report. 2006. Published by WWC and the Secretariat of the 4th World Water Forum. Accessed January 18, 2015. http://www.worldwatercouncil.org/fileadmin/world_water_council/documents/world_water_forum_4/Final_Report_4th_Forum.pdf.

Gies, Erica. 2009. "Is Water a Human Right or a Commodity?" *World Watch* 22(2): 22–27.

Goff, M., and B. Crow. 2014. "What is Water Equity? The Unfortunate Consequences of a Global Focus on 'Drinking Water.'" *Water International* 39(2): 159–171.

Harris, Leila M., Lucy Rodina, and Cynthia Morinville. 2015. "Revisiting the Human Right to Water from an Environmental Justice Lens." *Politics, Groups, and Identities* 3(4): 660–665.

Jordan, A., R. K. W. Wurzel, and A. Zito. 2005. "The Rise of 'New' Policy Instruments in Comparative Perspective: Has Governance Eclipsed Government?" *Political Studies* 53(3): 477–496.

Loftus, A. 2006. "Reification and the Dictatorship of the Water Meter." *Antipode* 38(5): 1023–1045.

Mehta, L. 2006. "Do Human Rights Make a Difference to Poor and Vulnerable People? Accountability and the Right to Water in South Africa." In *Rights, Resources and Accountability*, edited by P. Newell, pp. 63–78. London: Zed Books.

Mirosa, O., and L. M. Harris. 2012. "Human Right to Water: Contemporary Challenges and Contours of a Global Debate." *Antipode* 44(3): 932–949.

Parmar, P. 2008. "Revisiting the Human Right to Water." *The Australian Feminist Law Journal* 28: 77–96.

Peoples Water Forum Declaration-Istanbul. 2009. Accessed January 25, 2015. http://www.tni.org/article/peoples-water-forum-declaration.

Perera, V. 2014. "Engaged Universals and Community Economies: The (Human) Right to Water in Colombia." *Antipode* (early online view). https://doi.org/10.1111/anti.12097.

Scanlon, John, Angela Cassar, and Noemi Nemes. 2004. *Water as a Human Right?* IUCN Environmental Policy and Law Paper No. 51, IUCN: World Conservation Union.

Seventh World Water Forum 2nd Announcement. 2014. Accessed July 10, 2017. http://www.worldwatercouncil.org/fr/node/173.

Snitow, Alan, and Deborah Kaufman. 2008. "The New Corporate Threat to Our Water." In *Water Consciousness: How We All Have to Change to Protect Our Most Critical Resources*, edited by Tara Lohan, pp. 45–57. San Francisco, CA: AlterNet Books.

Subramaniam, Mangala. 2014. "Neoliberalism and Water Rights: Case of India." *Current Sociology* 62(3): 393–411.

Sultana, F., and A. Loftus. 2012. *Right to Water: Politics, Governance and Social Struggles*. New York: Earthscan.

4 CONTROLLING WATER RESOURCES FROM "ABOVE" ... 151

Synthesis of the Fourth World Water Forum. 2006. Published by WWC and the Secretariat of the 4th World Water Forum. Accessed January 18, 2015. http://www.worldwatercouncil.org/fileadmin/world_water_council/documents_old/World_Water_Forum/WWF4/synthesis_sept06.pdf.
Synthesis Report Seventh World Water Forum. 2015. Global Water towards Implementation. Accessed July 10, 2017. http://www.worldwatercouncil.org/sites/default/files/2017-09/SynthesisReport_7thWorldWaterForum_1.pdf.
Third World Water Forum Analysis. 2004. Published by WWC and the Secretariat of the 3rd World Water Forum. Accessed January 18, 2015. http://www.worldwatercouncil.org/fileadmin/world_water_council/documents/world_water_forum_3/3d_World_Water_Forum_analysis.pdf.
Third WWF Final Report. 2003. Published by Secretariat of the 3rd World Water Forum 2003. Accessed January 18, 2015. http://www.worldwatercouncil.org/fileadmin/world_water_council/documents/world_water_forum_3/3d_World_Water_Forum_FinalReport_BD.pdf.
United Nations. 2002. General Comment No. 15: The Right to Water (Articles 11 and 12 of the International Covenant on Economic, Social and Cultural Rights). Geneva: United Nations Economic and Social Council. Accessed March 16, 2011. http://www.unhchr.ch/tbs/doc.nsf/0/a5458d1d1bbd713fc1256cc400389e94?Opendocument.
UN-Water Decade Programme on Advocacy and Communication and Water Supply and Sanitation Collaborative Council. Undated. Accessed July 10, 2017. http://www.un.org/waterforlifedecade/pdf/human_right_to_water_and_sanitation_media_brief.pdf.
Venugopal, P. N. 2006. Coca Cola Moving Out of Plachimada? Accessed July 10, 2017. http://indiatogether.org/cokesaga-environment.
World Water Forum. 2000. Ministerial Declaration of The Hague on Water Security in the 21st Century. Accessed January 18, 2015. http://www.worldwatercouncil.org/fileadmin/world_water_council/documents/world_water_forum_2/The_Hague_Declaration.pdf.
World Water Forum. 2009. 5th World Water Forum Secretariat; Turkish Foreign Ministry Global Water Framework. Accessed January 25, 2015. http://www.worldwatercouncil.org/fileadmin/world_water_council/documents_old/World_Water_Forum/WWF5/global_water_framework_part_1_final.pdf—Part 1.
World Water Forum. 2012. "Access to Water for all, Supporting Local Communities' Access to Drinking Water and Sanitation." Report on the 6th World Water Forum Initiative. Published by WWC. Accessed January 25, 2014. http://www.worldwatercouncil.org/fileadmin/world_water_council/documents/official_documents/WWC_AccesstoWaterforAll_Report_June2014.pdf.

World Water Council. 2015. Final 7th World Water Forum Report. Accessed July 10, 2017. http://www.worldwatercouncil.org/sites/default/files/2017-09/Final%20Report_7th%20World%20Water%20Forum.pdf.

World Water Council Report. 2014. "6th World Water Forum 'Access to Water for All." Accessed January 25, 2014. http://www.worldwatercouncil.org/fileadmin/world_water_council/documents/official_documents/WWC_AccesstoWaterforAll_Report_June2014.pdf.

WWC Constitution & By-Laws. 2013–2015. Accessed January 25, 2015. http://www.worldwatercouncil.org/fileadmin/world_water_council/documents/Constitution_ByLaws_2013_2015.pdf (Also Available in French, Spanish and Russian).

WWC List of Members. April, 2014. Accessed January 25, 2015. http://www.worldwatercouncil.org/fileadmin/world_water_council/documents/wwc-membership/List_of_Members_per_college_April_2014.pdf.

WWC Water Crossroads. 2009. Accessed January 25, 2015. http://www.worldwatercouncil.org/fileadmin/world_water_council/documents_old/World_Water_Forum/WWF5/Water_at_a_Crossroad.pdf.

WWC World Water Vision Report (Presented by the World Commission on Water for the 21st Century). 2000. Accessed January 18, 2015. http://www.worldwatercouncil.org/library/archives/world-water-vision/vision-report/.

CHAPTER 5

Conclusion

I began this book by referring to the water crisis in Flint, Michigan and the discussion that emerged about the role of the state and private industry. Many have noted that communities struggle as groundwater levels deplete and the contamination of water adversely affects the health of the poor and marginalized more than others. Privatization disregards the role of communities in managing water resources. Proponents of privatization argue that a private company can be more efficient and deliver better service than a municipality. Governments can usually qualify for better interest rates on loans than corporations. Yet private corporations assert that they can deliver projects much more cost-effectively than a municipality because of their expertise and scale of work. In contrast, water activists, in local struggles and in the broader justice for water movement, say that private companies have no incentive to conserve water, because the more water they sell, the more money they make. Private companies are interested in people who can pay, making poor communities and/or those in rural areas vulnerable. And when water prices increase, these people resort to free, often contaminated, sources of water. These differences have been central to the challenges in villages and towns across the world, as well as in organized global protests such as the Alternative World Water Forum (AWWF) as discussed in this book, specifically in India and the United States.

There are four themes central to the analysis in this book: struggles to redefine the relations between the state, corporations, and other social actors around water; efforts to manage and coordinate water resources; local constitutive politics of inequalities—gender, class, and caste—among actors (people and state institutions); and the paradoxes of development initiatives because of the possibility of the destruction of environmental resources.

THE AMBIVALENT STATE IN THE NEXUS BETWEEN LOCAL STRUGGLES AND GLOBAL FORUMS

Protest, struggle, and the urge for equality are as old as constricting structures such as gender, caste hierarchy, inequality of power, wealth, and knowledge. Social movement theorists argue that movements, protests, and struggles are legitimate expressions of popular interests, and attempt to explain why, when, and how people protest and make claims. Protests and challenges to inequalities have been visible in discourse and movement activities all over the world. Efforts to challenge structural inequities also reveal the complex locations of different groups, particularly in the context of the current trends in globalization.

Globalization has had an impact on the economics and politics of states through complex interactive processes. Early conceptualizations of the state, such as that by Nettl (1968), sparked various arguments on the role and nature of the state. But much contemporary discussion on the state has been in the context of globalization. As discussed in Chapter 2, state responses have been conceptualized as representing retreat, centrism, and transformation. Discussions about the disprivileging of the state and its declining powers have been at the center of the literature on globalization. While the state may have lost some control to the dictates of global institutions, it has also acquired new areas of control, such as promoting privatization and competitiveness (Cable 1995). The state, in many countries, is active in creating and maintaining these conditions, and therefore to some degree, economic liberalization and globalization have gone hand in hand.

Economic liberalization, which is the neoliberal agenda, is equated with development—a process that includes market transactions and scarcity, among others, and that goes unhindered. The consolidation of neoliberalism in many regions of the world from the 1990s onwards makes the need for critiques more pressing, as does the growing awareness of the social and ecological costs of the decisions made by countries, such as India and Ecuador, to open themselves to the world's markets. The neoliberal language is also visible in the development discourse of "good governance," "partnership," and "ownership" (Escobar 2012). The development regime of the state as well as global financial institutions such as the World Bank and the IMF, along with NGOs should be seen as contributing to the creation of uneven geographies and inequalities within nations. It is important to recognize the tense co-existence of the economic and development policies at the level of the state, mostly guided by global financial institutions, on the one hand, and the ability of movements to problematize such policies from above and below and demand for change, on the other hand. This is clearly evident in countries across the world including India and the United States, the two countries I focus on in the analysis of the three cases of local struggles (see Chapter 3). The local struggles are not entirely separate from the global action to seek environmental justice.

A significant set of movements globally and across countries have challenged and continue to challenge the consequences of globalization and specifically the neoliberal agenda. The neoliberal order that is supported by powerful states and wealthy corporate interests has been expanding over time, but that order is also being vigorously challenged by movements acting both locally and transnationally. As Angelique Haugerud observes, "Neoliberalism has sparked a stunning array of popular countermovements" (2010: 112) that often target corporate and state power. Since the late 1990s, there has been a growing tendency to understand these kinds of movement politics as responding to various forms of "dispossession" unleashed as part of the latest wave of neoliberal globalization. Such endeavors for profit accumulation are closely linked to the global capitalist system (Harvey 2003). While neoliberalism itself is translocal in nature, movement dynamics, including their trajectory, can play out differently in different places (Subramaniam 2014). In other words, the local face and character of the consequences of neoliberalism and the complex intertwined relationship between local groups, NGOs, and alliances that enable ties (networks) across borders which metamorphose into global forums like the forum related to water (discussed in Chapter 4) are visible in different forms across places.

The relations between state institutions on the one hand, and local struggles and global forums on the other hand, is dynamic when it comes to adopting and implementing neoliberal tenets, particularly privatization. That is the relations have been flexible and have changed over time (Subramaniam 2015). As I have argued in this book, states have not evenly embraced neoliberalism or privatized social services completely. In fact, globalization processes, as discussed above, have on the one hand facilitated greater integration across borders, and on the other hand have also led states to assert their separate and independent status in the global economy. People's struggles and resistance to neoliberal tenets such as privatization may also lead to the reversal of decisions by state institutions at the national and sub-national levels. Therefore, the state itself must be conceptualized as comprising an apparatus of institutions at the local (such as the village) level and across provincial/districts or counties and at the national level. Such a conceptualization is useful for unraveling the dynamics of action across state institutions and with local, informal, or organized struggles. This is clearly evident in the case of the struggle in Plachimada.

The varying decision-making powers of state institutions—the *panchayat*, the Kerala state government, and the courts—have proved to be both a challenge and an opportunity. In the first instance, the Perumatty *panchayat* granted a license to Coca-Cola to set up a bottling plant in 2000. But in less than two years, the growing protests about the groundwater made the *panchayat* give a show-cause notice to Coca-Cola for details and then canceling the license. In a Public Interest Litigation (PIL), the *panchayat* also approached the Kerala State High Court, which ruled in favor of the protesting Adivasis. At the same time, the Court required the *panchayat* to permit Coca-Cola to continue operations. Having been legally advised to first comply with this order and then move the Supreme Court against it, the Gram Panchayat issued the permission for three months with 13 conditions.

As discussed in Chapter 3, the intense struggle and alliance with a national movement compelled the Kerala state government to order the closure of the plant. These dynamics across state institutions and with the protestors (including the Anti-Coca-Cola Struggle Committee) differ and have changed over time. The ambivalent state is constantly reshaping issues of control over water, leading to an uneven adoption of neoliberalism. This dynamic also raises questions about the state policy of participation in managing and maintaining water resources as the people of Plachimada had not been consulted in the establishment of the private factory. But the local panchayat responded to the protests and eventually pursued a policy that differed from the Kerala state government and the national government's agenda to attract and support private investment of all forms.

The Concerned Citizens' Coalition of Stockton (CCCS) also used the court system when it sued the City of Stockton for not conducting an environmental review of the contract as required by Californian law.

The City/City Council as a local state institution (similar to a *panchayat*) was targeted by the CCCS using legal means. The CCCS emerged victorious. Similarly, in the Alwar district of Rajasthan (India), local citizens have protested against Bisleri and Coke factories, and the litigation hearing is currently before the courts.

In all three cases of local struggles discussed in this book, the major actors include state institutions, activists, informal groups, and NGOs as well as the broader global water justice movement. As state institutions across the national and sub-national levels are targeted by local people's struggles and the global water justice movement, they also vary in terms of their response to the privatization agenda and the right to water. Thus, while national governments (and some NGOs) may participate in the World Water Forum (WWF), the state has not consistently adopted or implemented privatization, partly because of the fractionation of state institutions as well as the continuous and persistent movement action (local struggles and the AWWF). Such inconsistencies and contradictions make for what I call an *ambivalent* state. Therefore, the relations between state institutions, private water corporations, and people (particularly the marginalized) have involved both cooperation and discord.

COOPERATION AND DISCORD: RELATIONS BETWEEN THE STATE, CORPORATIONS, AND SOCIAL ACTORS

As noted by the UN Report on Water (2015), water management is the responsibility of many different decision-makers in the public and private sectors. The issue is how such shared responsibility can be turned into something constructive and elevated to a rallying point around which different stakeholders can gather and participate collectively to make informed decisions. International institutions, such as the World Bank and the World Trade Organization (WTO), promoting structural adjustments and trade liberalization have turned water into a prime battleground over the marketization and privatization of state-provided social services and public goods. Funding of the water infrastructure means attracting capital, and multinationals have entered this battleground in an attempt to gain profits and fulfill the goal of letting the market work. The goal of profit is in direct contradiction to the citizen's idea of water commons.

Today, we live in a world where everything is for sale. While most human societies and civilizations have traditionally recognized the water commons, our modern world is rapidly moving in the opposite direction by commodifying water as a product to be bought and sold in the marketplace. Instead of viewing water as a life-giving source, it is considered an economic good or commodity to be sold in the market. And once the market becomes the mechanism for the distribution of water, it is made available to those who have the ability to pay, rather than universally available to all people and nature. As discussed in Chapter 4, a major goal of the global water justice movement is to frame the notion of water commons as an important part of developing our water consciousness. It sets the parameters for deepening our understanding of why it is imperative to preserve and protect this resource.

Preserving and protecting water is about balancing local community control with relevant state (national and sub-national level) policy. If water really belongs to the people, for example, then local communities should be able to exercise ownership and control over their local watersheds. But such an understanding is distorted through the commodification and privatization of water. Under the guise of private-public partnerships, for-profit corporations are increasingly taking control of the management of water supplies and the delivery of water services in cities and towns. That is exactly what the Mayor of Stockton put in place in the city. As discussed in Chapter 3, Mayor Podesto even called for entering the twenty-first century in terms of the provision of services and to think of citizens as customers. Privatization of the source of water in the Alwar region was also initiated by the state and in this case the notion of partnership was not even considered.

Private corporations are very involved in the WWFs. As discussed in Chapter 4, the World Water Council (WWC) has direct links with two of the world's largest water corporations, Suez and Veolia. The president of the WWC is also president of a company equally owned by Veolia and a subsidiary of Suez. The WWF initially involved global financial institutions such as the World Bank, private water corporations, national governments, NGOs, and activists from around the world. But the AWWF was formed in 2003 mainly because of differences with the WWF in the understanding of the meaning of the right to water. The mission and goals and the organizations/groups involved in the AWWF as members and/or participants differ from those of the WWF based on the consideration of water as an economic good versus community-ownership and control.

The focus on community ownership and control is why the issue of water sovereignty is key to developing our water consciousness. Above all else, water is local. Local watersheds exist to serve the needs of people and nature in their surroundings. Governments can play an important role in managing and regulating water as a public or an ecological trust, but ultimately water sovereignty should reside in people and nature. Issues of control and management of water resources are closely connected to equity and politics of gender, caste, and class, as is evident from the analysis of the three cases discussed in Chapter 3.

WATER STRUGGLES AND MARGINALIZED POPULATIONS

The two main groups that are impacted by water availability and access are women and youth. Class, race, and/or caste should also be considered in understanding access to as well as control over water. Persistent poverty is usually the result of a vicious cycle in which limited income converges with limited access to resources. The global limit of ecological sustainability of water available for abstraction is reported to have been reached (Barker et al. 2000). Regionally, this limit has already been exceeded for about one-third of the human population and it will rise to about half by 2030 (WWAP 2012). Apart from access to sanitation and clean drinking water, the world's 850 million rural poor also lack access to water for agricultural production, which is usually their primary source of income (Soussan and Arriens 2004). Without access to improved agricultural water management, poverty in these regions will persist (Namara et al. 2010).

Water policies are often based on generalized perspectives that lack gender perspectives and local knowledge (WWAP 2012). By failing to integrate gender considerations into water resources management, gender inequities will persist, preventing the adoption of innovative solutions that may be put forward by women (WWAP 2012).

A daily struggle for water is one of the terrible burdens of poverty, especially for women and girls who spend endless hours fetching and carrying water over long distances. Sources of water are often unclean or unaffordable, or groups are simply cut off from using a particular water source. Many poor urban dwellers have to pay very high water prices to informal water vendors or do without water. Considering the factors of gender and class in accessing and controlling water, we should expect women and the poor to be actively involved in deliberations and decision-making regarding water, but that has not been the case.

As discussed in Chapter 2, the opportunity costs for participating may be steep for the poor. And yet it would be naïve to assume that this is the only reason for the lack of involvement of the poor in the process. Social power and local politics that are constitutive of water realities may constrain the participation of women and the poor or those from among lower caste categories. Women in the Alwar district region (in the case of TBS discussed in Chapter 3) who *attend* the water parliament sessions are rarely involved in the discussions or in addressing the conflicts. Their presence at these sessions is often incorrectly construed as participation. The gendered construction and process of conducting the water parliament sessions should be a concern for TBS, the local citizens, as well as activists such as Rajendra Singh (the "waterman") and policy makers.

Differing somewhat from the case of TBS, the Adivasis, who are a marginalized tribal population and who were deeply affected by the quality of water, challenged the *panchayat* to demand the shutdown of the Coca-Cola plant. Although farmers with land holdings in the area of the factory who were affected by the water shortage supported the movement, they did not participate in the protest activities initially; it was only when they experienced a reduction in water levels, making the borewells unusable that they altered their stance. Thus, the need for water drew in the relatively more privileged groups into the protests. Who gets to speak and engage in decision-making about managing water resources, including resolving conflicts regarding water, challenges homogeneous categorizations of peoples. Meyer (2008: 782) suggests that a post-exuberant approach to environmental politics should retain this "clear-eyed attention to the differential impact on human communities" and "arguments for restructuring social, economic, and political power" in order to achieve the justice and radical change that communities desire. What we need is a transformative vision of justice.

Who participates in discussions about the control over and management of water resources may also vary according to the setting/context, as well as whether residents are aware of concerns about water. The context of the protests in Plachimada and Alwar, particularly in a rural setting, differs from that of the CCCS in California. Because water supply systems in urban areas such as Stockton are generally not directly visible, many people may not realize that their water supply is being privatized under the cloak of efficiency. The small informal group of educated citizens of Stockton persisted in their protests against privatization, but there were no rallies or efforts to seek global alliances. As Lohan (2008) notes, citizens' protests against water privatization are growing in the United States, but they start small. Yet, the protests quickly expand to encompass issues of global economic justice, and their constituency grows to include people across lines of party, class, and race.

The discussions about who participates in deliberations of water-related matters are not confined to local struggles, but extend to global forums as well. The WWFs and AWWFs discussed in Chapter 4 draw our attention to the economic power of private corporations and global financial institutions in controlling water resources across the world. The WWF, which was first held in 1997, was intended to represent all stakeholders—private corporations, global financial institutions, NGOs, representatives of localized struggles, and activists. However, the issues raised and the emphasis on treating water as an economic good led to a split in the forums. The key idea in the differences among the stakeholders concerned the understanding of what constitutes right to water and who participates in making decisions regarding ownership and management of water resources. The first time that the issue of right to water was discussed in the WWFs was in 2006 in Mexico City. But the report of the fourth WWF in 2006 notes that the water debate is political as it is represented as both a basic right and an economic commodity. This discussion did not continue in the fifth WWF, but instead the sixth WWF in 2012 focused on people's need for water (access to water for all), acknowledging the importance of grassroots groups in addressing water challenges as well as recognizing the need for connecting local realities to global matters.

Several action projects were funded in the sixth WWF at Marseille, France. These key issues may suggest that the gap between the WWF and the AWWF is narrowing, but it is not as simple as that, because they do not reflect open dialogues between citizens and those who represent corporate interests. Moreover, the discussions in the WWF are typically in English, which limits the participation of activists and particularly representatives of citizens groups from across the world. Lack of resources precludes protestors and activists advocating for community or public ownership of water resources as a right to organize large AWWFs documenting their position or disseminating it widely as is done by WWFs. However, at the crux of the discord is the inconsistency in understanding the "rights" to water, and therefore the WWF and the AWWF differ in their perspectives on how to address the water crisis. The AWWF reflects the perspective of the local struggles such as those discussed in Chapter 3.

Water as a right (access to clean and safe water) is at the heart of the struggles such as in Plachimada. State policies enabling multinational investment were resisted by the local people of Plachimada as they collectively sought to control the use of water while trying to redefine the "water as a commodity" argument. In fact, the Plachimada Declaration notes that: "Water is not a commodity. We should resist all criminal attempts to marketise, privatise and corporatise water. Only through these means we can ensure the fundamental and inalienable right to water for the people all over the world" (Shiva n.d.: 15). This in itself speaks to the relevance of the need for the involvement of local residents/citizens in regulating the use of water.

The Paradoxes of Development and Justice

Discussions on the rights to water are also intertwined with issues of development and therefore to environmental justice. The concept of development is debated extensively across social science disciplines, but the approach adopted by international agencies and national governments points to the complexity of thinking of development. Often the approach is narrowly focused on investment and the creation of jobs, both of which are important and about which much is written. In fact, this approach is further conflated by the ambivalent state, as different state institutions alter their goals/policies temporally.

Consider, for instance, that in the case of Plachimada, the primary intention of both the local *panchayat* and the Kerala state government was that the Coca-Cola factory would provide jobs for the poor Adivasis, who had little say in the initial decision. Such an approach requires noting the differential power so evident in this and many environment-related struggles. It is mostly marginalized groups, such as the poor in the case of TBS, poor Adivasis in the case of Plachimada and ordinary citizens in the case of Stockton, who do not have formal political power to resist environmental destruction as well as the economic approach to all resources. Providing clean, accessible, affordable water is not only the most basic of all government services, but throughout history, control over water has defined the power structure of societies.

The power to define water as an economic good is tied to privatization in two ways: the first is by contracting water harvesting structures (such as setting up tube wells and constructing check dams) to private enterprises and the second is by providing private companies with control over water resources or supply systems within special schemes under state industrial policies intended to attract investment by connecting it to efficiency. The community protests against the privatization of the management of water resources and private sector industrialization also involve some contradictions that are rarely recognized or mentioned by activists. As water grows scarce and agriculture is adversely affected, village residents have increasingly sought jobs in local industries. Private companies such as Coca-Cola which are reported to indiscriminately draw on groundwater are also job providers, particularly for the local youth.

The lack of power for marginalized groups and ordinary citizens impacts what we refer to as environmental justice; the fair access to decision-making processes and resources as belonging to all (the community). Development initiatives across countries are often met with resistance because of the possibility of the destruction of environmental resources. Coca-Cola's dumping of sludge into paddy fields in Plachimada polluted the groundwater and is in direct opposition to the argument of development. These issues are similar to the concerns raised by other scholars (cf. Subramaniam and Zanotti 2015; Akiwumi 2015; Ahtuangaruak 2015 among others). The paradox of development and concerns of environmental justice have people/communities whose voices are suppressed on the one side and the ambivalent state on the other side. Local environmental struggles, such as the right to water (local and global), aim to end ecological devastation and economic exploitation. Power, justice, and rights are intertwined in the local struggles and global forums. People seek the right to self-determination and challenge current practices of corporations and state institutions. In addition, international conventions may be overlooked by countries when crafting policy (for example, UN protocols/agreements), as discussed in Chapter 4, and in doing so they sideline communities and people.

POLICY IMPLICATIONS

The impact of the neoliberal agenda on environmental justice has been devastating as people and landscapes are increasingly commodified and shaped by continual efforts to monetize the natural world (Smith 1984; Igoe 2010). Waves of neoliberal reforms have stimulated much debate about the value of nature, democratic processes, and the role of science, as concerned stakeholders, practitioners, and scholars seek practices that result in positive change (Meyer 2008: 782). Neoliberal agendas have also stimulated the decentralization and devolution of many forms of environmental management and policies from the state to municipal or local control. In some instances, the outcomes for justice have seemed positive, as the move toward more local control over management has fostered co-management regimes and cooperative community-conservation alliances (Charnley and Poe 2007). It has also fragmented and bureaucratized governance in a way that makes it increasingly hard for communities to address their justice concerns, thereby limiting rather than resolving ongoing environmental problems (Lemos and Agrawal 2006).

Institutions of governance lack authority to enforce the direct implementation of policy. This is because governance involves a wide variety of organizations that are responsible for different levels of oversight. These may be government agencies, statutory authorities, or third parties such as industry or community groups, which typically have varying levels of input in the decision-making process because of the differing levels of power they exert. This has led to tensions and contradictions amongst the various institutional entities and with nongovernance entities such as activists and movement struggles (both local and global). Global forums such as the WWF and the AWWF, combined with global action for water justice, have further complicated the role of state institutions across the national and sub-national levels. The unequal power exerted by institutional participants has made for shifts in relations across the various actors and the efforts of an ambivalent state to manage, coordinate, and enforce neoliberal policies have been erratic. All institutions and forums are confronting differences over how value is ascribed to water and the purpose of water governance. In these cases, as governance actors sought to improve access to water, they also had to confront crucial conflicts over who should hold the power and responsibility to manage this resource. This is evident in all the three cases discussed in Chapter 3. For instance, community efforts to revive water resources, such as the rejuvenation of rivers by the Alwar area communities (Rajasthan) with assistance from TBS, must be recognized as citizens being responsible, and therefore state control of such resources can prove controversial. Who should hold the power over water resources is also at the heart of the difference in the understanding of the right to water by the WWF and the AWWF.

The role of national governments in terms of considering the incorporation of the right to water in their constitutions is a critical aspect to consider in the efforts of the global movement for water justice. Several countries—Uruguay, South Africa, Ecuador, and many others—have passed right to water legislation and/or have incorporated it in their constitutions (see the discussion in Chapter 4). However, making policies and laws does not imply their adoption and enforcement, particularly because the government comprises a sprawling apparatus of institutions with varying levels of responsibility (and power) concerning water issues. So, the state has enabled and constrained people's/communities' right to water as it is wedged between global financial institutions and global forums on one side and community struggles on the other. This dynamic or ongoing dialogue with state institutions across the national and sub-national levels renders the state ambivalent. Discourses around the right to water and related policy tools have been problematic. But there has also been a growing acceptance of the importance of this right in policy agendas (e.g., the post-2015 Development Agenda).

Increasing competition for water and the ecological impacts of water management are significant policy issues around the world. The governance of water conflicts involves a "system of institutions, including rules, laws, regulations, policies, and social norms, and organizations involved in governing environmental resource use and/or protection" (Chaffin et al. 2014: 56). The effectiveness of governance arrangements to address growing conflicts over water and other resources is a critical question for research and practice. Realistically, the groups and organizations involved in protests (both local and global) are negotiated through formal and informal rules based on trust and relationships.

I caution against nostalgia for the centralized nation-state as well as against the celebration of local/local communities because both are complex terrains. The nation-state may be implicated in the process of marketization and globalization, but it is constantly being shaped by the multiplicities of challenges and struggles of local groups and global forums. Therefore, local communities and global water justice movements need to work "in and against" the state. While the emphasis on the local stems from the recognition of the need for decentralization of the management of water resources, it is not necessarily a space of commitment to justice because that space itself has its own hierarchy of power based on gender, caste, and class as well as other resources such as land ownership that enable maintaining the status quo.

Policy mechanisms related to water therefore need to balance the transformative vision of justice called for by Harvey (2000)—a vision that acknowledges the constraints in the current system, but also points to hopeful possibilities for equitable futures. Enabling such a vision requires long-term dialogues between state institutions, communities, and the broader global movement for water justice. These dialogues must incorporate considerations of recognitional and distributional issues that accompany the control over and management of resources.

The discussions in this book also have future implications for research. There is an urgent need to develop and adopt an interdisciplinary approach for understanding the strategies adopted by local struggles and national and transnational movements to seek justice (cf. Banerjee 2014). Moreover, successful mobilization strategies to demand justice are supported by "political alliances, social processes, economic pre-conditions, and cultural contexts" (Banerjee 2014: 817). Collaborations between social scientists, ecological and environmental engineers, as well as civil engineers are essential for studying how to address water justice by including the voices of citizens and communities. A synergy between the technical and the social aspects can be key to understanding issues related to water. This may also facilitate expanding their methods to include qualitative relationship building, in addition to quantitative methods, that gives voice to those who are less powerful in environmental justice contexts. Reliance on a single method can and has resulted in decisions that may continue to blame the victims and does not create the harmony and symbiosis that is sought by communities to address environmental justice issues.

References

Ahtuangaruak, Rosemary. 2015. "Broken Promises: The Future of Arctic Development and Elevating the Voices of Those Most Affected by It—Alaska Natives." *Politics, Groups, and Identities* 3: 673–677.
Akiwumi, Fenda A. 2015. "Analyzing Sierra Leone's Water Reform Efforts: Law, Environment, and Sociocultural Justice Issues." *Politics, Groups, and Identities* 3: 655–659.
Banerjee, Damayanti. 2014. "Towards an Integrative Framework for Environmental Justice Research." *Society and Natural Resources* 27: 805–819.
Barker, R., B. van Koppen, and T. Shah. 2000. *A Global Perspective on Water Carcity and Poverty: Achievements and Challenges for Water Resources Management.* Colombo: International Water Management Institute (IWMI).
Cable, Vincent. 1995. "The Diminished Nation-State: A Study in the Loss of Economic Power." *Daedalus* 124(2): 23–53.
Chaffin, B. C., H. Gosnell, and B. A. Cosens. 2014. "A Decade of Adaptive Governance Scholarship: Synthesis and Future Directions." *Ecology and Society* 19(3): 56. https://doi.org/10.5751/es-06824-190356.
Charnley, S., and Melissa Poe. 2007. "Community Forestry in Theory and Practice: Where Are We Now?" *Annual Review of Anthropology* 36: 301–336.

Escobar, Arturo. 2012. *Encountering Development: The Making and Unmaking of the Third World*. Princeton, NJ: Princeton University Press.
Harvey, David. 2000. *Spaces of Hope*. Berkeley: University of California Press.
Harvey, David. 2003. *The New Imperialism*. Oxford: Oxford University Press.
Haugerud, Angelique. 2010. "Neoliberalism, Satirical Protest, and the 2004 U.S. Presidential Campaign." In *Ethnographies of Neoliberalism*, edited by Carol Greenhouse, pp. 112–127. Philadelphia: University of Pennsylvania Press.
Igoe, Jim. 2010. "The Spectacle of Nature in the Global Economy of Appearances: Anthropological Engagements with the Spectacular Mediations of Transnational Conservation." *Critique of Anthropology* 30(4): 375–397.
Lemos, Maria Carmen, and Arun Agrawal. 2006. "Environmental Governance." *Annual Review of Environment and Resources* 31(1): 297–325.
Lohan, Tara (ed.). 2008. *Water Consciousness: How We All Have to Change to Protect Our Most Critical Resources*. San Francisco, CA: AlterNet Books and Watershed Media.
Meyer, John. 2008. "Comment: Political Theory and the Environment." In *The Oxford Handbook of Political Theory*, edited by John S. Dryzek, Bonnien Honig, and Anne Phillips, pp. 773–791. Oxford: Oxford University Press.
Namara, R. E., Hanjra, M. A., Castillo, G. E., Ravnborg, H. M., Smith, L., and Van Koppen, B. 2010. "Agricultural Water Management and Poverty Linkages". *Agricultural Water Management* 97(4): 520–527.
Nettl, J. P. 1968. "The State as a Conceptual Variable." *World Politics* 20(4): 559–592.
Smith, Neil. 1984. *Uneven Development: Nature, Capital, and the Production of Space*. Athens: University of Georgia Press.
Soussan, J., and Arriens, W. L. 2004. *Poverty and Water Security: Understanding How Water Affects the Poor*. Water for All Series No. 2. Asian Development Bank (ADB).
Subramaniam, Mangala. 2014. "Neoliberalism and Water Rights: Case of India." *Current Sociology* 62(3): 393–411.
Subramaniam, Mangala. 2015 "Introduction: States and Social Movements in the Modern World-System." *Journal of World Systems Research* (American Sociological Association's Political Economy of the World System Section Journal) 21(1): 1–7.
Subramaniam, Mangala, and Laura Zanotti. 2015. "Introductory Essay: Environmental Justice-Just Livelihoods." *Politics, Groups, Identities* 3(4): 649–654.
UN Report on Water. 2015. *Water for a Sustainable World*. Paris: UNESCO.
WWAP. 2012. *The United Nations World Water Development Report 4: Managing Water Under Uncertainty and Risk*. Paris: UNESCO.

Glossary

Adivasi an umbrella term for a heterogeneous set of ethnic and tribal groups considered to be the aboriginal population of India. They comprise a substantial indigenous minority of the population of India

Alternative World Water Forum (AWWF) formed as counter-forum to the World Water Forum. It represents people/citizens

Bharatiya Janata Party (BJP) a major national party in India promoting the "Hindu" ideology. The party won a majority (in coalition with other smaller political parties) at the national level and was in power for a brief period of two months in 1996, and then completed an entire five-year term between 1998 and 2004. The party currently holds power at the national level

Brahmins the highest of the traditional four varnas. They are seen as performing duties of worshipping Gods and holding administrative positions of power. They are landholders and are generally viewed as performing intellectual work. They are often described as the "dominant elites"

Chief Minister the elected leader of the party that gains the majority of seats in the state legislature. A Chief Minister heads a cabinet of ministers

Coca-Cola Virudha Samara Samithi (CCVS) the Anti-Coca-Cola People's Struggle Committee in Plachimada, Kerala (India)

Communist Party of India (CPI) the name of both the original party and its more pro-Moscow, moderate offshoot after the 1964 split

Communist Party of India (Marxist) (CPI (M)) name of the "left" of the two offshoots of the original CPI

Concerned Citizens' Coalition of Stockton (CCCS) formed in 2001 to monitor and challenge what it called Stockton City Mayor Gary Podesto's push for the privatization of the municipal water supplies

Congress (I) the Congress of Indira Gandhi, a political party formed from the erstwhile Indian National Congress

Dalit downtrodden. More than one-sixth of India's population (about 160 million people) are categorized as dalits. Under Articles 341 and 342 of the Indian Constitution, certain castes, specified by public notification, have been deemed to be Scheduled Castes. Castes in this category were and are still among the poorest sections of Indian society. Considered unclean and hence untouchable, and outside the pale of the caste system, the Scheduled Castes are subject to various types of discrimination ranging from physical avoidance to exclusion from Hindu temples. They are from among the lowest in the caste hierarchy. Other descriptive terms used to refer to them are "untouchables," "harijan" (a term coined by Gandhi), and dalits

Dharna a nonviolent, sit-in protest

District the smaller unit that states are divided into geographically and administratively. It can be seen as equivalent of a county in the United States

Jal Biradari a program to raise awareness about India's National Water Policy among people

Jal Bhagirathi Foundation (JBF), India established as a nonprofit organization on January 15, 2002 in response to the burgeoning water crises facing the Thar Desert in the state of Rajasthan, western India

Jal Suraksha, Adhikar, Mukti Declaration the National People's Water Forum Declaration. This forum comprised the Jal Swaraj Abhiyan and Rashtriya Jal Biradari

Johad traditional water harvesting structures or earthen check dams in the Rajasthan region to capture and conserve rainwater, which improve percolation and groundwater recharge

Lok Samitis community organizations of local village residents which participate in saving and managing water resources

Panchayat village-level governance institutions. It was traditionally a committee of five elite men ("panch" means "five"). The 1992, 73rd and 74th amendments to the Indian Constitution require the states of the Indian Union to enforce a local governance structure and provide a one-third reservation of the seats in these elected bodies to women

Panchayat Raj Institutions (PRIs) the governance system at the village, taluka, and district levels, particularly for the implementation of development programs. The system includes locally elected officials and bureaucracy

World Water Council (WWC) an international think tank based in Marseilles, France, which organizes the World Water Forum once every three years. Its goal is to raise awareness of global water issues

World Water Forum (WWF) the WWC membership sets the WWF's agenda. Critics of the WWF draw attention to the goals of those who came together to create the WWC in 1996

INDEX

A

Activist(s), 7, 14, 15, 17, 18, 38, 39, 47, 49, 67, 73–75, 82–85, 89, 91, 96, 98, 99, 103, 111, 117, 118, 123–125, 140, 153, 157, 158, 160, 161, 163, 164
Adivasi, 63, 68, 75, 80, 82, 83, 86, 91, 156, 160, 162
Agriculture/Agricultural, 3, 11, 66, 78, 86, 115, 123, 139, 143, 146, 159, 163
Alliance(s), 13, 34, 39, 43, 63, 69, 79, 80, 83, 114, 119, 125, 126, 155, 156, 160, 164, 166
  global, 34, 79, 160
  local, 34, 79
Alternative World Water Forum (AWWF), 10, 14, 15, 48, 96, 108, 110–118, 122, 124, 125, 141, 153, 157, 158, 161, 164
Ambivalent state, 9, 10, 16, 19, 33, 34, 48, 71, 85, 154, 156, 157, 162–165
Arvari, 16, 72, 73, 77, 78, 90

B

Beneficiaries, 37, 62
Bharatiya Janata Party (BJP), 61, 86
Bottled water, 5, 7, 8, 43, 44, 59, 60, 76, 84, 85, 121

C

Campaign, 5, 7, 14, 17, 42–44, 59, 60, 62, 64, 69, 75, 78–80, 82, 84, 86, 98, 103, 110, 111, 120, 121, 126
  signature, 79, 80, 82
Caste, 7, 19, 25, 28, 34, 35, 48–50, 70, 72, 74, 85, 89, 91, 154, 159, 160
Centre for Science and Environment (CSE), 73, 78, 84, 85
Citizens, 2, 4, 7, 9, 11, 13, 16, 17, 19, 28, 30, 34, 39, 40, 42, 47, 60, 62, 65–69, 71, 74, 77–82, 85, 90, 95, 98, 102, 106, 108–111, 115, 118, 122–124, 145–147, 156, 158, 160–164, 166
  forum, 102, 106, 108, 109

City Council, 43, 66, 70, 74, 78–82, 89, 157
Class, 1, 7, 8, 10, 17, 19, 25, 28, 34–36, 39, 45, 48, 70, 72, 74, 78, 80, 85, 89, 91, 124, 154, 159, 161
Coca-Cola, 12, 13, 17, 43, 60, 63, 67–69, 74–76, 78–80, 82–88, 91, 92, 104, 120, 121, 156, 160, 162, 163
Commodity, 7, 9, 28, 29, 40, 49, 76–78, 83, 96, 104, 111, 115, 117, 158, 161, 162
Communist Party of India (Marxist) (CPI(M)), 63, 86
Community, 1, 5, 7, 8, 12, 16, 18, 19, 25, 27, 28, 33–37, 41, 46–48, 59, 65, 67, 72, 73, 76–78, 81, 83, 84, 88, 90, 91, 98, 100, 107, 110, 117, 124, 125, 144, 146, 158, 161–165
Concerned Citizens' Coalition of Stockton (CCCS), 11, 13, 16, 17, 66, 69, 70, 74, 77–82, 84, 88, 89, 92, 156, 160
Consciousness, 37, 118, 119, 158
Constitution, 6, 49, 50, 99, 100, 106, 115, 123, 145–149, 164
Context, 26, 36, 47, 60, 61, 110, 154, 160, 166
Corporate, 10, 29, 42–44, 60, 66, 74, 82, 101, 108, 110, 113, 116, 120, 121, 155, 161
Corporation, 4, 6–8, 10, 14, 15, 18, 19, 27, 34, 38, 42, 43, 49, 50, 60, 64–66, 76, 78, 91, 95, 99–101, 105, 106, 110, 116, 118, 119, 123, 125, 153, 157, 158, 161, 163
 private, 49, 50, 66, 78, 91, 110, 118, 119, 125, 153, 157, 158, 161
 water, 60, 66, 95, 99, 100, 116, 125, 157, 158
Court, 13, 17, 68, 69, 75, 76, 79, 82, 84, 85, 87, 89, 123, 124, 156, 157
 High, 17, 68, 75, 84, 87, 88, 124, 156
 Superior, 69, 89, 123
Covenant, 96, 98, 99, 110, 125, 127

D
Dalit, 49, 50
Declaration, 43, 76, 83, 96, 97, 101, 103–109, 113–118, 124, 126, 127, 162
 Ministerial, 102–107, 109
Development, 3, 5, 13, 27, 33, 37, 38, 44, 60, 63, 70, 85, 86, 88, 97, 100–106, 109, 116, 119, 120, 127, 142–149, 154, 155, 162, 163, 165
Dharna, 68, 69, 79, 80, 83

E
Economic good, 40, 50, 75, 110, 158, 161, 162
Environment, 1, 2, 5, 9–11, 14, 16, 18, 47–50, 62, 64, 73–75, 78, 79, 82, 83, 85, 86, 88, 89, 96, 99, 101, 110, 111, 114, 115, 120, 123, 127, 144–149, 154–156, 160, 162–166

F
Farmer(s), 3, 36, 44, 46, 67, 72, 83, 86, 89, 91, 160
Flint, MI, 1, 153
Free market, 28

G
Gandhi, 50
Gender, 7, 8, 10, 17, 19, 25, 28, 34, 35, 45, 47, 48, 64, 70, 72, 74, 85, 89, 91, 103, 109, 118, 154, 159

General Comment No. 15, 97, 98, 125
Global Forums, 9, 10, 15, 17, 19, 26, 39, 48, 60, 91, 95, 99, 154, 155, 161, 163–165
Globalization, 7–10, 15, 16, 25–28, 31–35, 38, 43, 48, 49, 83, 90, 101, 102, 115, 154–156, 165
Government, 1, 9, 11–13, 16–18, 28–32, 39, 41, 42, 44, 50, 60–63, 65–67, 69, 72, 73, 75, 76, 78, 85–88, 90, 91, 97, 98, 100, 105, 106, 109–113, 115, 116, 120–123, 125, 147, 156–158, 162, 164, 165
  delegation, 103, 105
  state, 13, 16, 17, 63, 65, 75, 87, 105, 156
Grassroots, 13, 14, 17, 19, 35–38, 64, 69, 79, 80, 84, 108, 161
Groundwater, 3, 11, 60, 67, 68, 72, 76, 78, 80, 84, 87, 91, 118, 120, 121, 123, 153, 163

I
India, 3, 4, 6, 11, 12, 16, 17, 26, 33–35, 37, 40, 41, 43, 46, 50, 60–64, 66–68, 78, 79, 83–85, 91, 92, 104, 109, 117, 120, 121, 123, 146, 153, 154, 157
Indigenous, 7, 30, 41, 42, 44, 46, 50, 70, 71, 77, 79, 98, 103, 104, 117, 124, 126
  knowledge, 7, 70, 77–79
  people, 44, 50, 71, 77, 98, 103, 104, 117
International Monetary Fund (IMF), 28, 29, 34, 49, 77, 90, 113, 155
Istanbul, Turkey, 14, 101, 102, 104, 105, 107, 108, 112–114, 131, 132, 140, 141
Italy, 96, 111, 112, 122
  Florence, 14, 96, 111, 112

J
Jal Bhagirathi Foundation (JBF), India, 109, 111
Janata Dal, 86, 87
Japan, 102, 103, 146
  Kyoto, 101–103, 106, 111
Johad, 11, 67, 78
Justice, 7, 11, 15, 27, 47–50, 62, 86, 88, 96, 98, 99, 106, 110, 113, 114, 118, 120, 124, 125, 149, 153, 155, 157, 158, 160–166
  environmental, 11, 47, 49, 50, 62, 86
  recognitional, 47, 165

K
Kerala, 12, 13, 17, 43, 60, 63, 64, 67, 68, 74–76, 83, 84, 86–88, 91, 121, 156, 162
Kerala State Pollution Control Board (KSPCB), 74, 87, 88
Knowledge, 2, 9, 10, 33, 47, 77–80, 102, 117, 126, 154, 159
  water, 47, 77
Korea, 14, 101, 102, 105, 108, 109, 112, 116, 124, 136, 146
  Daegu and Gyeongbuk, 101, 102, 108, 112

L
Language, 15, 19, 46, 48, 102, 105, 118, 123, 154
Law/legal, 5, 6, 13, 14, 17, 25, 31, 40, 42, 43, 50, 59, 62, 68, 69, 79, 80, 82, 84, 88, 89, 91, 97, 100, 103, 106, 107, 110, 116, 117, 119, 123, 144–146, 148, 156, 157, 165
Liberalization, 9, 29, 49, 61, 90, 154, 157
  economic, 61

176 INDEX

Livelihood, 3, 45, 46, 71, 98, 120, 144, 145
Lok Samitis, 72

M
Marginalized, 1, 2, 10, 14, 36–38, 47, 79, 98, 99, 124, 153, 157, 160, 162, 163
Market, 6, 8, 10, 16, 28, 29, 31, 33, 34, 41–43, 46, 61, 64–66, 72, 90, 92, 95, 110, 124, 154, 157, 158
Marseille, France, 14, 18, 95, 96, 100–102, 105, 112–114, 116
Mayor, 13, 16, 66, 69, 70, 74, 79–82, 84, 88, 89, 158
Member(s), 13, 18, 19, 36, 70, 72, 74, 79, 81, 83, 92, 95–98, 100, 101, 106, 111, 113, 124
Mexico, 3, 14, 101–103, 107, 111–114, 124, 129, 147, 161
  Mexico City, 14, 101–103, 107, 111–114, 161
Mobilize (mobilizing), 9–12, 16, 17, 30, 34, 38, 39, 47, 67, 71–73, 84, 96, 98, 113
Morocco, 102, 104
  Marrakesh, 101, 102, 106
Municipality, 153

N
Neoliberal/Neoliberalism, 2, 5, 9, 10, 16, 19, 25–34, 38, 48–50, 75, 85, 90, 154, 155, 163, 164
The Netherlands, 102, 123, 147
  The Hague, 101–103, 106, 111
Network(s), 9, 10, 12–17, 20, 27, 28, 30, 38, 39, 48, 49, 68, 92, 96, 98–100, 111, 114, 118, 119, 122, 155

Non-governmental organizations (NGOs), 9, 11, 14, 17–19, 26, 27, 30, 34, 36, 37, 39, 41, 63, 64, 80, 83, 85, 89, 90, 97, 103–105, 111, 117, 118, 122, 125, 155, 157, 158, 161

O
OMI/Thames, 16, 69, 79, 81, 82, 84, 89
Organization(s), 7, 9, 11, 12, 14, 15, 17, 18, 20, 26, 27, 31, 35–37, 39–41, 43, 49, 66, 67, 72–75, 83, 96, 100, 101, 103–105, 108–110, 113, 115–121, 124, 144, 146, 158, 164, 165
  grassroots, 35, 36, 38, 108
  informal, 15, 17, 73
Ownership, 6, 9, 10, 16, 18, 34, 41, 42, 44, 45, 63–65, 70, 89, 92, 95, 96, 117, 147, 155, 158, 161

P
Panchayat, 12, 13, 17, 19, 60, 68, 69, 73, 75, 76, 79, 80, 87, 88, 91, 92, 117, 156, 160, 162
Pani sansad (water parliament), 41, 79, 80
Parliamentarian(s), 104, 105, 107, 114
Participation, 8, 11, 37, 41, 44–47, 61, 71–73, 75, 91, 106, 108, 111, 113, 118, 126, 145, 156, 160, 161
People's Water Forum (PWF), 9, 10, 14, 17, 48, 96, 111–114, 117, 125
Pepsi/PepsiCo, 43, 85, 88, 92, 121
Perumatty, 12, 68, 87, 91, 156
Plachimada, 11–13, 15, 17, 43, 60, 64, 66–68, 74–80, 82–84, 86–88, 91, 104, 118–121, 156, 160–163

INDEX 177

strategies, 79
Podesto, 13, 69, 81, 82, 88, 89, 158
Political parties, 86, 88, 113
Pollution, 2, 12, 68, 80, 83, 86, 88, 104, 115, 145, 147
Privatization, 1, 2, 5, 7–11, 13–19, 25, 28–30, 33, 34, 39, 40, 42–44, 47–49, 59–61, 64–67, 69–71, 74–79, 81, 82, 84–86, 88–91, 95–98, 103, 104, 107, 108, 111, 112, 114, 116–118, 122, 123, 125, 153–158, 160, 162
Protest(s), 5, 7, 10, 12, 13, 15–17, 26, 34, 35, 38, 39, 43, 47, 50, 59–61, 64, 69, 74–76, 79, 83, 85–87, 111, 112, 119, 120, 122, 153, 154, 156, 160, 162, 165
Public–private, 42, 70, 77, 96, 107

R
Rainwater harvesting, 7, 78, 84, 121
Rajasthan, 11, 12, 15, 16, 41, 42, 63, 64, 66, 67, 72, 78, 91, 109, 157, 164
Rajendra Singh, 11, 12, 66, 67, 82, 85, 109, 160
Rally, 68, 121
Resources, 1–5, 7, 8, 10, 11, 14–17, 19, 25–31, 33–35, 38, 41, 43–46, 48, 49, 59–72, 74–77, 79–81, 85, 86, 88, 89, 91, 96, 98, 100, 101, 108, 110–112, 114–116, 118, 121, 122, 125, 142–149, 153, 154, 156, 159–165
Right, 1, 4–8, 11, 15, 17–19, 25, 26, 28–31, 33, 37, 40–45, 47–50, 69–72, 76, 77, 83, 87, 91, 95–99, 102–119, 122–127, 136, 139, 140, 144–149, 157, 158, 161–165
 economic, 107, 125, 161
 human, 8, 15, 40, 42, 43, 48–50, 69, 77, 83, 96–98, 103, 106–110, 112, 113, 115, 116, 122, 123, 125, 127, 144
 social, 40, 91, 97, 107, 125
Rupees, 72, 75, 86, 92
Rural, 9, 11, 12, 14, 17, 18, 36, 37, 47, 50, 67, 76, 77, 96, 111, 153, 159, 160

S
State, 4, 6–12, 15–19, 25–41, 43, 45, 48–50, 60–64, 66–71, 73–76, 78–80, 82, 83, 85–88, 90–92, 96, 98, 99, 102–106, 109, 110, 113, 117–119, 121–125, 145–149, 153–158, 162–165
 centrist, 31, 32, 154
 retreat, 31, 32, 154
 transformation, 31, 32, 154
Stockton, 13–16, 42, 60, 66, 67, 69, 70, 74, 79, 81, 82, 84, 88, 89, 110, 118, 156, 158, 160, 162
Suez, 14, 18, 66, 95, 100, 101, 123, 158

T
Tarun Bharat Sangh (TBS), 11, 12, 15–17, 41, 46, 60, 66, 67, 72, 73, 77, 79, 80, 118, 160, 162
 strategies, 79
Transnational, 2, 7–10, 14, 16, 18, 19, 26, 39, 43, 112, 114
Turkey, 102, 104, 111–113, 148
 Istanbul, 102, 104, 112

U
United Nations, 2–4, 18, 40, 45–47, 71, 96–100, 105–110, 113, 124–126, 163
United Nations Universal Declaration of Human Rights, 96, 126

United States/US, 3, 8, 11, 26, 35, 37, 38, 42–44, 48, 50, 60, 61, 64–66, 72, 79, 84, 85, 90, 92, 108, 149, 153, 155, 160
User groups, 46, 62

V
Veolia, 14, 18, 66, 95, 101, 116, 158
Village(s), 4, 6, 11–13, 15, 16, 19, 33, 37, 39, 41, 47, 50, 64, 67, 68, 71–73, 75, 78, 80, 83, 89–92, 102, 103, 117, 118, 121, 153, 156, 163

W
Water Age, 4–6, 59
  commons, 68, 77, 116, 118, 157, 158
  first, 4, 59
  second, 4, 5, 59, 60
  third, 4–6, 59
Woman/Women, 10, 11, 35–37, 46, 47, 68, 73, 80, 86, 118, 127, 159, 160
World Bank, 2, 9, 14, 18, 28, 29, 34, 49, 64, 77, 90, 95, 97, 99, 100, 112–114, 123, 155, 157, 158
World Trade Organization (WTO), 9, 28, 29, 49, 90, 113, 157
World Water Council (WWC), 9, 14, 18, 95–97, 99–103, 105–110, 112, 114, 158
World Water Forum (WWF), 9, 14, 17, 18, 48, 91, 95–97, 99–113, 116, 118, 122, 124, 125, 157, 158, 161
World Water Vision, 101–103, 106

The manufacturer's authorised representative in the EU is Springer Nature Customer Service Centre GmbH, Europaplatz 3, 69115 Heidelberg, Germany. If you have any concerns regarding our products, please contact ProductSafety@springernature.com

Printed and bound by CPI Group (UK) Ltd, Croydon, CR0 4YY

23/03/2026

02076735-0001